1990

The IMA Volumes
in Mathematics
and Its Applications

Volume 16

Series Editors
Avner Friedman   Willard Miller, Jr.

Institute for Mathematics and
Its Applications
IMA

The **Institute for Mathematics and its Applications** was established by a grant from the National Science Foundation to the University of Minnesota in 1982. The IMA seeks to encourage the development and study of fresh mathematical concepts and questions of concern to the other sciences by bringing together mathematicians and scientists from diverse fields in an atmosphere that will stimulate discussion and collaboration.

The IMA Volumes are intended to involve the broader scientific community in this process.

Avner Friedman, Director
Willard Miller, Jr., Associate Director

IMA PROGRAMS

1982-1983	**Statistical and Continuum Approaches to Phase Transition**
1983-1984	**Mathematical Models for the Economics of Decentralized Resource Allocation**
1984-1985	**Continuum Physics and Partial Differential Equations**
1985-1986	**Stochastic Differential Equations and Their Applications**
1986-1987	**Scientific Computation**
1987-1988	**Applied Combinatorics**
1988-1989	**Nonlinear Waves**
1989-1990	**Dynamical Systems and Their Applications**

SPRINGER LECTURE NOTES FROM THE IMA:

The Mathematics and Physics of Disordered Media
Editors: Barry Hughes and Barry Ninham
(Lecture Notes in Math., Volume 1035, 1983)

Orienting Polymers
Editor: J.L. Ericksen
(Lecture Notes in Math., Volume 1063, 1984)

New Perspectives in Thermodynamics
Editor: James Serrin
(Springer-Verlag, 1986)

Models of Economic Dynamics
Editor: Hugo Sonnenschein
(Lecture Notes in Econ., Volume 264, 1986)

Avner Friedman

Mathematics in Industrial Problems

With 64 Illustrations

Springer-Verlag
New York Berlin Heidelberg
London Paris Tokyo

Avner Friedman
Institute for Mathematics and Its Applications
University of Minnesota
Minneapolis, MN 55455, USA

Mathematics Subject Classification (1980): 05C20, 35L65, 49A21, 49A29, 65C20, 73D05, 73D25, 73H10, 76B10, 76D05, 76F99, 76T05, 78A45, 90B10, 93C22, 94B35

Library of Congress Cataloging-in-Publication Data
Friedman, Avner.
    Mathematics in industrial problems / Avner Friedman.
        p.   cm. — (The IMA volumes in mathematics and its
    applications ; v. 16)
        Bibliography: p.
        ISBN 0-387-96860-1
        1. Engineering mathematics.   I. Title.   II. Series.
    TA330.F75   1988
    620′.0042—dc19   88-24909

Camera-ready copy prepared by the author using T_EX.
Printed and bound by R.R. Donnelley & Sons, Harrisonburg, Virginia.
Printed in the United States of America.

9 8 7 6 5 4 3 2 1

ISBN 0-387-96860-1 Springer-Verlag New York Berlin Heidelberg
ISBN 3-540-96860-1 Springer-Verlag Berlin Heidelberg New York

The IMA Volumes
in Mathematics and Its Applications

Current Volumes:

Volume 1: Homogenization and Effective Moduli of Materials and Media
 Editors: Jerry Ericksen, David Kinderlehrer, Robert Kohn, J.-L. Lions

Volume 2: Oscillation Theory, Computation, and Methods of Compensated Compactness
 Editors: Constantine Dafermos, Jerry Ericksen,
 David Kinderlehrer, Marshall Slemrod

Volume 3: Metastability and Incompletely Posed Problems
 Editors: Stuart Antman, Jerry Ericksen, David Kinderlehrer, Ingo Müller

Volume 4: Dynamical Problems in Continuum Physics
 Editors: Jerry Bona, Constantine Dafermos, Jerry Ericksen, David Kinderlehrer

Volume 5: Theory and Applications of Liquid Crystals
 Editors: Jerry Ericksen and David Kinderlehrer

Volume 6: Amorphous Polymers and Non-Newtonian Fluids
 Editors: Constantine Dafermos, Jerry Ericksen, David Kinderlehrer

Volume 7: Random Media
 Editor: George Papanicolaou

Volume 8: Percolation Theory and Ergodic Theory of Infinite Particle Systems
 Editor: Harry Kesten

Volume 9: Hydrodynamic Behavior and Interacting Particle Systems
 Editor: George Papanicolaou

Volume 10: Stochastic Differential Systems, Stochastic Control Theory and Applications
 Editors: Wendell Fleming and Pierre-Louis Lions

Volume 11: Numerical Simulation in Oil Recovery
 Editor: Mary Fanett Wheeler

Volume 12: Computational Fluid Dynamics and Reacting Gas Flows
 Editors: Bjorn Engquist, M. Luskin, Andrew Majda

Volume 13: Numerical Algorithms for Parallel Computer Architectures
Editor: Martin H. Schultz

Volume 14: Mathematical Aspects of Scientific Software
Editor: J.R. Rice

Volume 15: Mathematical Frontiers in Computational Chemical Physics
Editor: D. Truhlar

Forthcoming Volumes:

1987-1988: *Applied Combinatorics*
 Applications of Combinatorics and Graph Theory to the Biological and Social Sciences
 q-Series and Partitions
 Invariant Theory and Tableaux
 Coding Theory and Applications
 Design Theory and Applications

Contents

Mathematics in Industrial Problems

Introduction

Building a bridge between mathematicians and industry is both a challenging task and a valuable goal for the Institute for Mathematics and its Applications (IMA). The rationale for the existence of the IMA is to encourage interaction between mathematicians and scientists who use mathematics. Some of this interaction should evolve around industrial problems which mathematicians may be able to solve in "real time." Both Industry and Mathematics benefit: Industry, by increase of mathematical knowledge and ideas brought to bear upon their concerns, and Mathematics, through the infusion of exciting new problems.

In the past ten months I have visited numerous industries and national laboratories, and met with several hundred scientists to discuss mathematical questions which arise in specific industrial problems. Many of the problems have special features which existing mathematical theories do not encompass; such problems may open new directions for research. However, I have encountered a substantial number of problems to which mathematicians should be able to contribute by providing either rigorous proofs or formal arguments.

The majority of scientists with whom I met were engineers, physicists, chemists, applied mathematicians and computer scientists. I have found them eager to share their problems with the mathematical community. Often their only recourse with a problem is to "put it on the computer." However, further insight could be gained by mathematical analysis.

Some of the scientists have presented their problems in an IMA Seminar on Industrial Problems. The book is based on the questions raised in that seminar and subsequent discussions. Each chapter is devoted to one of the talks. The chapters are self–contained. Reference to the relevant mathematical literature appears usually at the end of each chapter. Each chapter also contains problems for mathematicians, most of which are of interest to industrial scientists. In some cases the problems have been partially or completely solved; in these instances an outline of the solution is presented.

The speakers in the industrial seminar have given us at IMA hours of delight and discovery. My thanks to Robert Ore (UNISYS), Michael Honig (Bellcore), Stephan Kistler (3M), Peter Castro (Eastman Kodak), Young–Hwa Kim (3M), Roger Hastings (UNISYS), Brian Flannery (Exxon),

4

Michael Ward (Caltec and IBM), John Ockendon (Oxford University), Craig Poling (Honeywell), Willard L. Miranker (IBM), David Ross (Eastman Kodak), D. George Wilson (Oak Ridge National Laboratories), Fan Chung (Bellcore), Blaise Morton (Honeywell), Louis Romero (Sandia National Laboratories), James McKenna (AT&T Bell Laboratories), Edward Bissett (General Motors), Tom Haigh (Honeywell), Luis Reyna (IBM), Bhashrpilla Gopinath (Bellcore), and Alan Cox (Honeywell).

Superb assistance was given by Patricia V. Brick and Kaye Smith who typed the manuscript and Stephen Mooney who drew the figures. Thanks are also due to Bob, Ceil, Mary, Kelly, Susan, Deb and Leslie of the IMA staff for creating the environment in which we all thrive. Finally, I am grateful to Willard Miller, Jr. who has been a source of encouragement throughout this industrial adventure.

Avner Friedman
Director
Institute for Mathematics
and its Applications
June 20, 1988

1

Scattering by Stripe Grating

When bismuth is substituted at certain lattice points of rare earth garnet crystals, a material is formed which exhibits an almost ideal magnetic stripe domain structure. The structure consists of a one–dimensional periodic array of regions of nearly constants magnetization, so that the magnitudes of the magnetization vectors in two adjacent regions are the same, but their directions are opposite. This condition produces the alternating–sign magnetic permeability tensors shown in section 1.7(3). When a beam of light is passed through such a film, it splits into many diffracted beams, all sharing a plane defined by the periodicity direction (i.e., the y-component of all the wave vectors is the same; see Figure 1.1). The angles of the diffracted–beam directions relative to the incident–beam direction are determined by the periodicity length.

By appropriate tuning of the material parameters and film thickness one can achieve high–efficiency beam deflection. That is, it is possible to channel most of the incident energy into a single beam which is at a specific angle to the incident beam.

It is an important property of these garnet films that the application of a magnetic field can modify the stripe domain structure. It is possible in this way to rotate the direction of periodicity and to change the periodicity length. Thus, one has the ability to vary both the azimuthal and the polar deflection angles purely electronically (i.e., without mechanically modifying the grating). Furthermore, by applying time–varying magnetic fields such changes can be modulated at high frequencies (limited only by the ability of the domain structure to respond).

The applications of such a variable–angle light deflector are many. One which has been studied at UNISYS is a multiple–output optical switch, in which a single input laser beam can be directed, by application of appropriate magnetic fields, to any one of many possible output directions. Such a switch could be an important component, for example, of a fiber–optic communications network.

On October 9, 1987 Robert Ore from UNISYS made initial presentation of the the diffraction model. After subsequent discussion with Walter Littman and Peter Rejto (from the School of Mathematics of the University of Minnesota) he has completed the model for the simpler case of diffraction of particles in an electric field. The material presented in sections 1.1–1.6 was communicated to me by Robert Ore on December 17, 1987.

1.1   The Physical Problem

We are given an infinite slab of thickness d with potential energy $V(x)$, periodic in x of period L and constant in y, z; $V(x) = V_0$ if $0 < x < \dfrac{L}{2}$ and $V(x) = -V_0$ if $\dfrac{L}{2} < x < L$ (V_0 positive); see Figure 1.1.

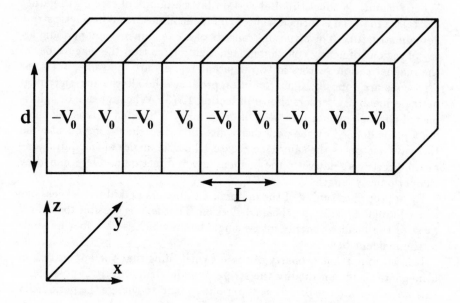

FIGURE 1.1.

A beam of particles of energy $E = q^2$ are incident on the slab from below with wave vector $\vec{q}$; $|\vec{q}| = q$.

Problem. What is the wave function $\Psi_{\vec{q}}(\vec{r})$ for the stationary problem, for $|z| \gg d$ (i.e., far below and far above the slab)?

1.2   Relation to the Time-dependent Problem

One should think of the wave function as time-dependent function $\Psi(\vec{r}, t)$ satisfying the Schrödinger equation

$$i\frac{\partial}{\partial t}\Psi + \Delta\Psi - V\Psi = 0 \tag{1.1}$$

in the whole space, such that

as $t \to -\infty$, $\Psi(x,t)$ tends to the free solution

(1.2)

of (1.1)   (i.e., to the solution with $V \equiv 0$).

In the present case, the free solution we use in the plane wave with wave vector $\vec{q}$:

$$\Phi_{\vec{q}}(\vec{r},t) = \Phi_{\vec{q}}(\vec{r})e^{-iEt} \quad \text{with} \quad \Phi_{\vec{q}}(\vec{r}) = e^{i\vec{q}\cdot\vec{r}} .$$

The wave function $\Psi \equiv \Psi_{\vec{q}}$ then satisfies the integral equation

$$\Psi_{\vec{q}}(\vec{r},t) = \Phi_{\vec{q}}(\vec{r},t) - i \int\limits_{-\infty}^{t} dt' \int d^3r' \quad G^{(0)}(\vec{r} - \vec{r}';t - t')$$

(1.3)

$$\cdot V(\vec{r}')\Psi_{\vec{q}}(\vec{r},t')$$

where $G^{(0)}$ is the free Schrödinger Green's function:

$$(i\frac{\partial}{\partial t} + \Delta_{\vec{r}}) \quad G^{(0)}(\vec{r} - \vec{r}';t - t') = i\delta^3(\vec{r} - \vec{r}')\delta(t - t'),$$

$$G^{(0)}(\vec{r} - \vec{r}';t - t') = 0 \quad if \quad t < t'.$$

Since V is time-independent one can show that

$$\Psi_{\vec{q}}(\vec{r},t) = \Psi_{\vec{q}}(\vec{r})e^{-iEt}$$

(1.4)

where $\Psi_{\vec{q}}(\vec{r})$ satisfies the Lippmann–Schwinger equation

$$\Psi_{\vec{q}} = \Phi_{\vec{q}} + G^{(0)}_{\vec{q}} * V\Psi_{\vec{q}}$$

(1.5)

with

$$G^{(0)}_{\vec{q}} (\vec{r} - \vec{r}') = \lim_{\epsilon \downarrow 0} \int \frac{d^3k}{(2\pi)^3} \quad \frac{e^{i\vec{k}\cdot(\vec{r}-\vec{r}')}}{\vec{q}^2 - \vec{k}^2 + i\epsilon} = -\frac{1}{4\pi} \frac{e^{i|\vec{q}| \, |\vec{r}-\vec{r}'|}}{|\vec{r} - \vec{r}'|}.$$

(1.6)

The function $\Psi_{\vec{q}}(\vec{r})$ determined by (1.5) is the stationary solution which we seek to study in section 1.1. In sections 1.3–1.5 we shall construct $\Psi_q(\vec{r})$ by a method of superposition.

1.3   Form of Solutions for $|\mathbf{z}| > \mathbf{d}$

We can write superposition of free solutions in the form

$$\Psi_q(\vec{r}) = \begin{cases} e^{i\vec{q}\cdot\vec{r}} + \displaystyle\sum_{n=-\infty}^{\infty} R_n e^{i\vec{q}_n^- \cdot \vec{r}} & , \quad z < -\dfrac{d}{2} \\ \displaystyle\sum_{n=-\infty}^{\infty} T_n \, e^{i\vec{q}_n^+ \cdot \vec{r}} & , \quad z > \dfrac{d}{2} \end{cases} \tag{1.7}$$

where $\vec{q} = (q_x, q_y, q_z)$ and

$$\vec{q}_n^{\pm} = (q_x + \frac{2\pi n}{L} \, , \, q_y, \pm\sqrt{\vec{q}^{\,2} - (q_x + \frac{2\pi n}{L})^2 - q_y^2}) \quad ; \tag{1.8}$$

the R_n are reflection coefficients and the T_n are the transmission coefficients. Only the terms with

$$(q_x + \frac{2\pi n}{L})^2 + q_y^2 \leq \vec{q}^{\,2} \quad , \quad n = 0, \pm 1, \pm 2, \ldots \tag{1.9}$$

are significant for $|z| >> d$; the other terms in (1.7) decay exponentially as $|z| \to \infty$.

Notice that $\vec{q}_0^+$ is the direction of the beam of particles incident to the slab.

1.4   Form of Solutions Inside the Slab

Since $V(x)$ is periodic, by Bloch's theorem (see reference in section 1.6) every solution Ψ_p in the slab is given by

$$\Psi_p(\vec{r}) = e^{2\pi i p x / L} \, u_p(\vec{r})$$

where $-\frac{1}{2} < p \leq \frac{1}{2}$ and $u_p(\vec{r})$ is periodic in x of period L. Consider the solution in the "unit cell" shown in Figure 1.2.

Since

$$u_p(L, y, z) = u_p(0, y, z), \quad \frac{\partial}{\partial x} u_p(L, y, z) = \frac{\partial}{\partial x} u_p(0, y, z),$$

we get

$$\Psi_p(L, y, z) \quad = e^{2\pi i p} \Psi_p(0, y, z) \, ,$$

$$\tfrac{\partial}{\partial x} \Psi_p(L, y, z) \quad = e^{2\pi i p} \tfrac{\partial}{\partial x} \Psi_p(0, y, z) \, .$$

We solve for $\Psi_p(\vec{r})$ by finding the most general solution for $0 \leq x \leq L/2$ and for $\frac{L}{2} \leq x \leq L$ and matching them at $x = \frac{L}{2}$ and $x = 0, L$:

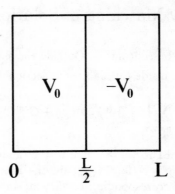

FIGURE 1.2.

$$\Psi_p(x;y,z) = e^{ik_y y + ik_z z} F_p(x) \ ,$$

$$F_p(x) = \begin{cases} A\, e^{ik_x^+ x} + B e^{-ik_x^+ x} \ , & 0 \le x \le \dfrac{L}{2} \\[2mm] C\, e^{ik_x^- x} + D e^{-ik_x^- x} \ , & \dfrac{L}{2} \le x \le L \end{cases}$$

where $k_x^{\pm} = \sqrt{\vec{q}^{\,2} - k_y^2 - k_z^2 \mp V_0}$. There are three parameters on which Ψ_p depends: p, k_y, k_z. Imposing continuity of Ψ_p and $\dfrac{\partial}{\partial x}\Psi_p$ yields four linear homogeneous equations for A, B, C, D. The vanishing determinant of this system then gives a transcendental equation for p, k_y, k_z. Noting that, when the solution inside the slab is matched to that outside, the x and y components of the wave vectors must match, we find that $k_y = q_y$ and $p = \dfrac{Lq_x}{2\pi}$ mod 1, so only k_z remains as a variable; thus, it remains to solve the transcendental equation for k_z.

The equation has infinitely many solutions. For $t = V_0/q^2 \longrightarrow 0$, they may be characterized as follows.

$$\begin{cases} k_z^{\pm,m} = \sqrt{q^2 - k_y^2 - (\frac{2\pi}{L})^2\,(m \pm p)^2} \ , & m = 1, 2, \ldots \\[2mm] k_z^0 = \sqrt{q^2 - k_y^2 - (\frac{2\pi}{L})^2\,p^2} \ . \end{cases} \qquad (1.10)$$

Thus, for $m \le N$, there are $2N + 1$ solutions. Since the equation for k_z only depends on k_z^2, there is a similar set of solutions with $k_z < 0$.

These solutions we call *modes*. Including $k_z > 0$ and $k_z < 0$, then, there are $2(2N + 1)$ modes in the slab for $m \le N$.

1.5   Boundary Matching of Solutions

At $z = \pm\dfrac{d}{2}$, Ψ must be continuous. Outside the slab the solution is given as in section 1.3 and, inside the slab, as in section 1.4, i.e.,

$$\Psi_{in}(\vec{r}) = \sum_{\eta,m}\left[C^{\eta,m}\,\Psi_{p,\eta,m}(\vec{r}) + C^{r,\eta,m}\,\Psi^r_{p,\eta,m}(\vec{r})\right]$$

where $\Psi_{p,\eta,m}$ ($\eta = \pm1$ for $m \geq 1$ and $\eta = 1$ (a single mode) for $m = 0$) are the modes for $k_z^{\eta,m} > 0$, $\Psi^r_{p,\eta,m}$ are the modes for $k_z^{\eta,m} < 0$, and $C^{\eta,m}, C^{r,\eta,m}$ are the corresponding coefficients in the superposition; "r" stands for reflection. We must express Ψ_{in} in terms of Bloch's Theorem decomposition:

$$\Psi_{in} = e^{2\pi ipx/L}\,u_{p,in}$$

where $u_{p,in}$ is a periodic function in x of period L. We can write the Fourier series

$$u_{p,in}(x,y,z) = \sum_{n=-\infty}^{\infty} e^{2\pi inx/L}\,\tilde{u}^n_{p,in}(y,z)$$

where the coefficients $\tilde{u}^n_{p,in}$ can be expressed in terms of $u_{p,in}$, or Ψ_{in}, by integration.

We may now match the solution inside and outside the slab. There are two matching conditions for each Fourier component, for each surface. If $n \leq N$, we then have a total of $4(2N + 1)$ equations.

There are $2N + 1$ coefficients $C^{\eta,m}, 2N + 1$ coefficients $C^{r,\eta,m}, 2N + 1$ reflection coefficients R_n and $2N + 1$ transmission coefficients T_n, for a total of $4(2N + 1)$ coefficients. Thus the total number of equations equals the number of unknowns, and the system of equations is inhomogeneous. It follows that the matching problem is in principle solvable.

1.6   Remarks and References

For the time-dependent problem described in section 1.2, see

[1] L.F. Schiff, Quantum Mechanics, McGraw–Hill, New York, 1968, pp. 298–320,

[2] B.A. Lippmann and J. Schwinger, Phys. Rev. 79, p. 469 (1950).

The diffraction model described in section 1.1 is known as the Quantum Mechanical Kronig–Penney Diffraction Model; see

[3] C.A. Wert and R.M. Thomson, Physics of Solids, McGraw-Hill, New York, 1970, pp. 361–368.

For details on Bloch's theorem see [3; pp. 361–368].

A typical diffraction calculation is the "N–beam dynamical" model in which it is assumed that N $R'_n s$ and N $T'_n s$ are non-zero and all the remaining coefficients vanish; see

[4] J.M. Cowley, Diffraction Physics, North–Holland–American Elesvier, New York, 1975.

Assuming that R_0, $R_{\pm 1}$ and T_0, $T_{\pm 1}$ are non-zero and all other coefficients vanish, Robert Ore finds that the method of section 1.5 is in excellent agreement with the 3–beam model as long as $V_0 L^2$ is small enough.

1.7 Mathematical Issues

(1) Establish rigorously the method of superposition described in sections 1.3–1.5. This means that one has to:

   (i) prove that the transcendental equation for k_z has infinite number of solutions $k_z^{\pm,m}$;

   (ii) Study the asymptotic behavior of $k_z^{\pm,m}$ as $m \to \infty$;

   (iii) establish the matching of the series expansions about $z = \pm d/2$, by showing that the linear equations for the uknown coefficients are uniquely solvable;

   (iv) prove convergence of the formal series.

Even for small $\epsilon = V_0/q^2$ this has not been carried out.

(2) Study the asymptotic behavior of the solution as $|z| \to \infty$ (after solving problem (1)).

(3) The Lippmann-Schwinger integral equation (1.5) has been used to establish existence of a solution with short-range potential, i.e., with potential $V = V(\vec{r})$ which is $O(|\vec{r}|^{-2-\epsilon})$ as $|\vec{r}| \to \infty$ ($\epsilon > 0$); see

[5] T. Ikebe, Eigenfunctional expansion associated with Schrödinger operators and their applications to scattering theory, Archive Rat. Mech. Anal., 5 (1960), 1–34.

For the present potential, find whether (1.5) has a solution, and if so,

   (i) study the asymptotic behavior of the solution as $|z| \to \infty$;

   (ii) does the solution coincide with the solution obtained by the method of superposition.

Robert Ore was actually initially interested in solving an optical diffraction by magnetic stripe grating, a problem similar to that described in section 1.1 but involving the Maxwell equations. In the "unit cell" instead of $V = \pm V_0$ we have the "magnetic permeability tensor"

$$\overset{\leftrightarrow}{m}_\pm = \begin{pmatrix} 1 & \mp i\epsilon & 0 \\ \pm i\epsilon & 1 & 0 \\ 0 & 0 & 1 \end{pmatrix} \quad , \quad \epsilon \text{ constant,}$$

and Maxwell's equations in the slab are

$$\overset{\rightarrow}{\nabla} \times \left[\overset{\leftrightarrow}{m}_\pm^{-1} \cdot (\overset{\rightarrow}{\nabla} \times \vec{A}) \right] - n^2 q^2 \vec{A} = 0 ,$$

$$\overset{\rightarrow}{\nabla} \cdot \vec{A} = 0$$

where

$$\vec{A} = \text{ vector potential (Coulomb gauge)},$$
$$q = \text{ wave number } = \tfrac{\omega}{c},$$
$$n = \text{ index of refraction } (n > 1),$$

and

$$\vec{B} = \overset{\rightarrow}{\nabla} \times \vec{A}, \quad \vec{E} = -\frac{\partial}{\partial t} \vec{A} = i\omega \vec{A}, \quad \vec{H} = \overset{\leftrightarrow}{m}_\pm^{-1} \cdot \overset{\rightarrow}{\nabla} \times \vec{A} .$$

(4) Extend the Kronig-Penney diffraction model to the optimal diffraction by a magnetic stripe grating.

Scattering models for diffraction gratings for the wave equation are described in

[6] C.H. Wilcox, Scattering Theory for Diffraction Gratings, Springer-Verlag, New York, 1984.

The geometry is given in Figure 1.3

with periodic boundary in the x–direction; the wave equation is satisfied on one side of the curved boundary, and the boundary condition is

$$\frac{\partial u}{\partial \nu} = 0 \quad \text{(acoustically hard boundary)}$$

or

$$u = 0 \quad \text{(accoustically soft boundary)}.$$

The corresponding model for the Maxwell equations is discussed in:

[7] Electromagnetic Theory of Gr gs, R. Petit, editor, Topics in Current Physics, vol. 22, Springer–Verlag, Berlin, 1980.

For both models it would be interesting to study the asymptotic behavior as $z \to \infty$.

FIGURE 1.3.

1.8   Partial Solution to Problem (3)

H. Bellout and A. Friedman have studied the integral equation (1.5) and proved that if $V_0 d$ is sufficiently small (depending on k, L) then there exists a unique bounded solution $\Psi_{\vec{q}}$, and the transmission and reflection coefficients can be computed.

2

Packing Problems in Data Communications

On October 16, 1987 Michael Honig from Bell Communications Research (Bellcore) reported on ongoing research in signal processing and posed some open problems. The presentation in sections 2.1–2.3 is based on his lecture notes and preprint [1].

2.1   Motivation and Problem Statement

Currently, nearly all voice and data traffic over the public switched telephone network is supported by twisted pair copper wires, which connects customer premises to the central office. Although eventually these copper wires will be replaced by high–speed fiber optic links, for at least the next few years the introduction of any new services over the public switched telephone network will have to be supported by existing copper facilities. It is therefore of interest to determine the maximum rate at which information can be communicated over such a channel.

A digital communications system consists of an information source, a transmitter, channel, and receiver. The information source generates a message picked from a set having N elements. For example, the message can be represented as a sequence of n bits. The transmitter converts the message to a time function $u(t)$, which is the input to a channel. Although our main interest is in the copper wire channel, the channel might also be the atmosphere, a fiber optic link or a magnetic recording device. We assume that the channel can be modeled as a linear operator, i.e.,

$$y(t) = \int_0^t h(t-s)u(s)\,ds \qquad (2.1)$$

where $u(t)$ and $y(t)$ are the input and output of the channel, respectively, and $h(t)$ is the impulse response of the channel. It is assumed that $h(t) = 0$, $t < 0$. The receiver attempts to estimate the transmitted sequence of bits given the channel output, $y(t)$.

Let $u_1(t), \ldots, u_N(t)$ denote an *input set* and $y_1(t), \ldots, y_N(t)$ denote an *output set* of the channel, respectively, and assume that the receiver must be able to distinguish among the N possible outputs in the time interval $[0, T]$. Define the *minimum distance* between outputs in the L_p sense over the interval $[0, T]$ as

$$d_{\min}(N, T) = \min_{i \neq j} \|y_i - y_j\|_{p,T} . \tag{2.2}$$

In the following problem statements it is assumed that all inputs in the input set have bounded L_q norm over the interval $[0, T]$, i.e., $\|u_j\|_{q,T} \leq 1$, $1 \leq j \leq N$.

(P1) Given T and N, find the input set that maximizes $d_{\min}(N, T)$.

(P2) Given $\delta > 0$ and T, find the input set such that $d_{\min}(N, T) \geq \delta$, with N as large as possible. Let $N_{\max}(T)$ denote the largest possible N.

(P3) Given $\delta > 0$ and N, find the input set that minimizes the time T such that $d_{\min}(T, N) \geq \delta$. Let $T_{\min}(N, \delta)$ denote the minimum time.

It is apparent that

$$T_{\min}(N, \delta) = \inf \left\{ T \mid N_{\max}(T, \delta) \geq N \right\} \tag{2.3}$$

and

$$N_{\max}(T, \delta) = \max \left\{ N \mid T_{\min}(N, \delta) \leq T \right\}. \tag{2.4}$$

For a given δ and h, the *maximum channel throughout* is defined as

$$MCT(\delta) = \sup_{T>0} \frac{\log N_{\max}(T, \delta)}{T} . \tag{2.5}$$

For $p = q = 2$, this quantity has been referred to as ϵ capacity [2].

The particular values of p and q chosen depend on the assumed channel model. For example, a realistic constraint for the copper wire channel is that the inputs to the channel be constrained in amplitude, which implies that $q = \infty$. In other cases of practical interest, however, it may desirable to restrict the energy of the input, in which case $q = 2$.

The appropriate value of p depends on the type of noise, or random disturbance, which the channel and receiver may introduce. For example, we may assume that any random disturbance, which the channel may introduce, is masked by the finite precision with which the receiver can measure the channel output. If the receiver can measure the channel output to within a finite accuracy of $\pm \delta/2$, then to communicate reliably, the outputs of the channel should be separated in amplitude by δ. An alternative interpretation of this situation is that the channel (and receiver) add noise, about which all we know is that it is bounded by $\delta/2$, and that the receiver can sample the channel output with infinite precision. The appropriate value of p in this case is ∞. The case $p = q = \infty$ is of special interest for the copper wire channel.

The case $p = 2$ is relevant if the channel is modeled as a linear operator followed by a white Gaussian noise source (i.e., thermal noise), and it is assumed that the receiver computes a maximum likelihood estimate of the input message given the received signal, $r(t)$, over the time interval $[0, T]$. The maximum likelihood estimate of the transmitted message is the j that minimizes $\|r(t) - y_j(t)\|_{2,T}$ where $1 \leq j \leq N$. In this case it is desirable that the output set be separated as much as possible in the L_2 sense.

Of course, variations of (P1)–(P3) are also of interest. For example, in some situations it may be desirable to impose both a maximum amplitude (L_∞) and energy (L_2) constraint on the inputs and ask that the L_2 (or L_∞) distance between outputs be maximized.

2.2 $p = q = \infty$

Consider first the special case $h(t) = Ae^{-\alpha t}$ where $A > 0, \alpha > 0$.

Theorem 2.1 *[1]. Let $h(t) = Ae^{-\alpha t}$ $(A > 0, \alpha > 0), N = 2^k$. Denote by $\bar{b}_{j\ell}$ the ℓ-th digit in the binary expansion of an integer j, $0 \leq j \leq N - 1$; set $b_{j\ell} = 1$ if $\bar{b}_{j\ell} = 1$ and $b_{j\ell} = -1$ if $\bar{b}_{j\ell} = 0$. Given $d > 0$, a set of inputs u_j which minimize T such that*

$$\|h * u_i - h * u_j\|_{\infty, T} \geq d$$

is given by

$$u_j(t) = b_{j\ell} \quad if \quad (\ell - 1)\Delta < t < \ell\Delta \qquad (1 \leq \ell \leq k)$$

where

$$\Delta = -\frac{1}{\alpha} \, \log(1 - \frac{d}{2A}) \ .$$

Theorem 2.2 *[1]. In case $p = \infty$,*

$$\sup_{T>0} \frac{\log N_{\max}(T, d)}{T} = \lim_{T \to \infty} \frac{\log N_{\max}(T, d)}{T} \ .$$

This number is called the *maximum channel throughout* and is denoted by $MCT(d)$; it measures bits per second (*bps*).

From Theorem 2.1 it readily follows that

$$\text{MCT}\,(d) = -\frac{\alpha}{\log(1 - \frac{d}{2A})} \quad if \quad h(t) = Ae^{-\alpha t} \ . \tag{2.6}$$

One important aspect of the general packing problem is to estimate MCT for general h. The following estimates are derived in [1]:

$$\frac{1}{r(d)} \leq \text{MCT}\,(d) \leq \frac{2K}{d} \tag{2.7}$$

where

$$K = \int_0^\infty |\frac{dh}{dt}| \, dt \ , \qquad \int_0^{r(d)} |h(t)| \, dt = \frac{d}{2} \ .$$

For any $\tau > 0, \alpha$ real, the *spread* of the channel $, \sigma(\tau, \alpha)$, is defined as $\sigma(\tau, \alpha) = \sigma^+(\tau, \alpha) - \sigma^-(\tau, \alpha)$ where

$$\sigma^+(\tau, \alpha) = \max_u \int_{-\infty}^{\tau} h(\tau - s)u(s) \, ds \ ,$$

$$\sigma^-(\tau, \alpha) = \min_u \int_{-\infty}^{\tau} h(\tau - s)u(s) \, ds,$$

and u varies in the class

$$\left\{ -1 \leq u(s) \leq 1 \quad \text{for} \quad -\infty \leq s \leq \tau, \quad \int_{-\infty}^{0} h(-s)u(s) \, ds = \alpha \right\}.$$

One can show that for any signal $y(t)$ with $y(t_0) = \alpha$,

$$\sigma^-(\tau, \alpha) \leq y(t_0 + \tau) - y(t_0) \leq \sigma^+(\tau, \alpha) \ .$$

This suggests that one may be able to estimate $MCT(d)$ in terms of the spread function. In fact, if τ^* is defined by

$$\sup_{\alpha} \sigma(\tau^*, \alpha) = d \tag{2.8}$$

then (by a paper in preparation by L. Honig, K. Steiglitz, S. Boyd and B. Gopinath)

$$\text{MCT}(d) \leq \frac{1}{\tau^*} \ . \tag{2.9}$$

Problems. (1) Study properties of $\sigma^{\pm}(\tau, \alpha)$ and evaluate $\sigma(\tau, \alpha)$.

(2) Find sharp upper estimates on $MCT(d)$

2.3   The Case $p = q = 2$

Assume that $h(t) = 0$ if $t < 0$ or $t > \tau$. Then if H is the operator of convolving with h then $H : L_2(0, T) \longrightarrow L_2(0, T + \tau)$, its adjoint H^* maps $L_2(0, T+\tau)$ into $L_2(0, T)$, and $H^* H$ is a compact self adjoint operator from $L_2(0, T)$ into $L_2(0, T)$. Denote by (ϕ_j, λ_j) a complete sequence of eigenfunctions and eigenvectors of $H^* H$.

For any input $u \in L_2(0, T)$ there is an expansion

$$u(t) = \sum_{i=1}^{\infty} a_i \, \phi_i(t) \ .$$

The corresponding output $y(t)$ is given by $Hu = \Sigma a_i H \phi_i$. Setting

$$\psi_i = \frac{H \, \phi_i}{\lambda_i^{1/2}},$$

it is easily seen that the ψ_i form an orthonormal sequence in $L^2(0, T + \tau)$ and

$$y(t) = \sum_{i=1}^{\infty} b_i \psi_i(t) \quad , \quad b_i = a_i \, \lambda_i^{1/2} \; .$$

We can identify $u \leftrightarrow \{a_i\}, y \leftrightarrow \{b_i\}$. Then, if $\|u\|_{2,T} = 1$,

$$\sum_{i=1}^{\infty} \frac{b_i^2}{\lambda_i} = 1, \qquad (2.10)$$

that is, y lies on the elliposoid $\sum_{i=1}^{\infty} \frac{x_i^2}{\lambda_i} = 1$.

The problem of packing N signals in time $T + \tau$ reduces to choosing sequences $\{a_i^k\}$ or sequences $\{b_i^k\}$ $(b_i^k = a_i^k \lambda_i^{1/2})$ satisfying (3.1) such that

$$\sum_{i=1}^{\infty} \frac{(b_i^k - b_i^\ell)^2}{\lambda_i} \leq d \qquad \forall \, k \neq \ell; \;\; 1 \leq k, \ell \leq N.$$

If the λ_i are monotonically decreasing then one can pack a maximal number N of outputs $y_j = \{b_i^j\}$ for which only this first N coordinates are possibly non-zero. Thus the packing problem reduces to packing N points in an $N - 1$ dimensional ellisposoid with axes $\lambda_1^{1/2}, \ldots, \lambda_N^{1/2}$. This problem is open for $N \geq 4$. For bounds on MCT see [2].

2.4   Solution to the Spread Problem

A. Friedman and M.L. Honig have recently studied properties of the spread function. They proved:

Theorem 2.3 *If $h(t) \neq 0$ a.e. and*

$$meas \left\{ \frac{h(\tau + t)}{h(t)} = \mu \right\} = 0 \qquad \forall \, \tau > 0, \; \mu \in \mathbb{R} \; ,$$

then the optimal control which maximizes $\sigma^+(\tau, \alpha)$ is given by

$$u_0(s) = \begin{cases} \operatorname{sgn} h(-s) & \text{if } \frac{h(\tau - s)}{h(-s)} > \lambda \\ -\operatorname{sgn} h(-s) & \text{if } \frac{h(\tau - s)}{h(\tau - s)} < \lambda \end{cases}$$

where λ is uniquely determined by

$$\int_{\left\{ \frac{h(\tau-s)}{h(-s)} > \lambda \right\}} |h(-s)| \, ds - \int_{\left\{ \frac{h(\tau-s)}{h(-s)} < \lambda \right\}} |h(-s)| \, ds = \alpha \; .$$

In particular, if $h(t) > 0$ and $\log h(t)$ is convex, then

$$u_0(s) = \begin{cases} 1 & \text{if} & -\infty < s < \mu \\ -1 & \text{if} & \mu < s < 0 \end{cases}$$

where μ is determined by

$$\int_{-\infty}^{\mu} h(-s)\, ds - \int_{\mu}^{0} h(-s)\, ds = \alpha \; ;$$

if $h(t) > 0$ and $\log h(t)$ is concave, then

$$u_0(s) = \begin{cases} -1 & \text{if} & -\infty < s < \tilde{\mu} \\ 1 & \text{if} & \tilde{\mu} < s < 0 \end{cases}$$

where $\tilde{\mu}$ is determined by

$$- \int_{-\infty}^{\tilde{\mu}} h(-s)\, ds + \int_{\tilde{\mu}}^{0} h(-s)\, ds = \alpha.$$

The proof of Theorem 2.3 is obtained by performing perturbations of the optimal control.

Using Theorem 2.3 they computed $\sigma^{\pm}(\tau, \alpha)$ for functions such as

$$h(t) = \Sigma\, a_i\, e^{-\beta_i t} \qquad (a_i > 0,\ \beta_i > 0),$$
$$h(t) = e^{-\mu t}\, \cos \omega t \qquad (\mu > 0,\ \omega > 0).$$

Theorem 2.4 *For h as in Theorem 2.3,*

$$\frac{\partial}{\partial \alpha}\, \sigma^{\pm}(\tau, \alpha) > 0 \; .$$

For $h > 0$ with $\log h$ convex or $\log h$ concave

$$\frac{\partial}{\partial \tau}\, \sigma^{\pm}(\tau, \alpha) > 0.$$

It was pointed out by Stephen Boyd that $\sigma(\tau, \alpha)$ is concave in α, and $\partial \sigma(\tau, \alpha)/\partial \alpha \leq 0$ if $\alpha \geq 0$.

2.5   REFERENCES

[1] M.L. Honig, S. Boyd, B. Gopinath and E. Rantapaa *On optimum signal sets for digital communications with finite precision and amplitude constraints.* Bell Communications Research Technical Memorandum TM –ARH–008–026, 1987.

[2] W.L. Root *Estimates on ϵ-capacity for certain linear communication channels* IEEE, Transactions on Information Theory, Vol. IT–14, May 1968, pp. 361–369.

3

Unresolved Mathematical Issues in Coating Flow Mechanics

A coating flow is a fluid flow that is used for covering a surface area. It is associated, for example, with application of paint by brush or roller, withdrawing a flat sheet or web immersed in paint bath, passing a web between two rollers, one of which has previously been coated with the coating fluid, etc. Examples of coating flows used in industry are diagrammed in Figure 3.1. In rimming flow a limited amount of liquid is put inside a tube or cylinder to be coated, which is oriented horizontally and spun about its axis so that the liquid forms a continuous film covering the outer rim. In dip coating there is the complication of upstream inflow and downstream outflow. In the other coating flows additional complications arise due to contact lines in the film–forming zones; the flow boundary comes in in contact with the fixed boundary, and the behavior of the flow lines near such contact points is not fully understood. This topic remains an active area of research in coating flows.

For a general introduction on coating we refer to the books of Middleman [1] and Pearson [2] and a review article by Ruschak [3]. Here we shall deal only with Newtonian flows. In section 3.1 we describe a coating problem which is common in industry; the presentation is based on a talk given by Stephan Kistler from 3M on October 30, 1987, and on a paper by Kistler and Scriven [4]. In section 3.2 we give a brief review of the mathematical literature. Simplified models are described in section 3.3, and possible future directions are indicated in section 3.4.

3.1 Curtain Coating

The equations of fluid flow are the conservation of momentum

$$Re\left[\frac{\partial \vec{v}}{\partial t} + (\vec{v} \cdot \nabla) \vec{v}\right] - \nabla \cdot T = St \, \vec{f} \qquad (3.1)$$

and the continuity equation

$$\nabla \cdot \vec{v} = 0 \quad ; \qquad (3.2)$$

here Re and St are the Reynolds and Stokes numbers, $\vec{v}$ is the velocity vector and T is the stress tensor; $\vec{f}$ is a unit vector in the direction of the

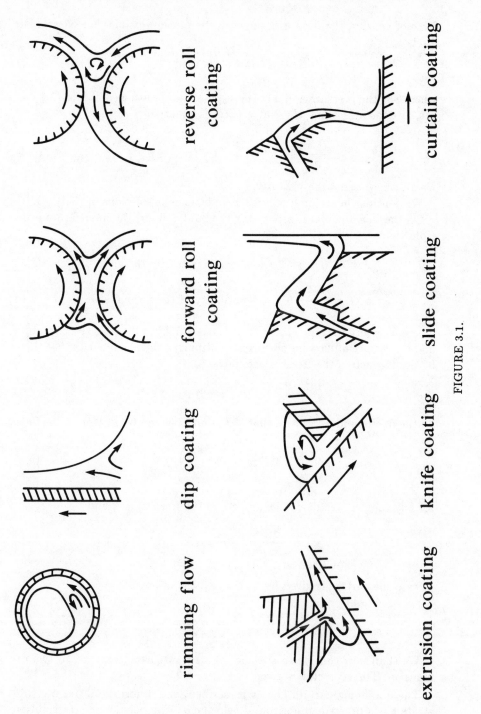

reverse roll coating

curtain coating

forward roll coating

slide coating

dip coating

knife coating

rimming flow

extrusion coating

FIGURE 3.1.

force of gravity; the fluid is assumed to be incompressible. For Newtonian fluids

$$T_{ij} = -\delta_{ij}p + \left(\frac{\partial v_i}{\partial x_j} + \frac{\partial v_j}{\partial x_i} \right) \qquad (3.3)$$

where p is the pressure and $\vec{v} = (v_1, v_2, v_3)$. Substituting (3.3) into (3.1) we get, after some scaling, the Navier–Stokes equation

$$\frac{\partial \vec{v}}{\partial t} + (\vec{v} \cdot \nabla) \vec{v} - \nu \nabla^2 \vec{v} + \nabla p = \vec{f} \qquad (3.4)$$

where ν represents the viscosity.

We consider henceforth only time–independent solutions of (3.4), (3.2).

On the free boundary (separating the fluid and air) the momentum principle yields

$$\vec{\tau} \cdot T \vec{n} = 0 \qquad \text{(no shear exterted by the air)} \quad, \qquad (3.5)$$

$$\vec{n} \cdot T \vec{n} = \frac{1}{Ca} K \qquad \text{(continuity of the normal stress)} \qquad (3.6)$$

where $\vec{\tau}$ is any tangent vector, $\vec{n}$ is the normal vector, K is the curvature and Ca is the capillary number. The continuity principle requries that there be no flow across the free surface, i.e.,

$$\vec{n} \cdot \vec{v} = 0 \qquad (3.7)$$

on the free boundary. Notice that (3.5) and (3.6) can be written in the form

$$\sum_{i,j} \tau_i \left(\frac{\partial v_i}{\partial x_j} + \frac{\partial v_j}{\partial x_i} \right) n_j = 0 \qquad (3.8)$$

for any tangent vector $\vec{\tau} = (\tau_1, \tau_2, \tau_3)$ to the free surface, and

$$\frac{1}{2} \sum_{i,j} n_i \frac{\partial v_i}{\partial x_j} n_j - p = \alpha K \qquad (\alpha = \frac{1}{Ca}) \qquad (3.9)$$

in case the surface is given by $x_3 = \phi(x_1, x_2)$, where

$$K = \nabla \cdot \frac{\nabla \phi}{\sqrt{1 + |\nabla \phi|^2}} \quad.$$

We shall now consider a specific problem of curtain coating, in two dimension. The geometry is given in Figure 3.2.

The impinging curtain flow is prescribed, and a horizontal moving substrate has a prescribed horizontal velocity U. The behavior of the solution as $x \to \infty$ approximates the uniform flow as described in the "outflow"

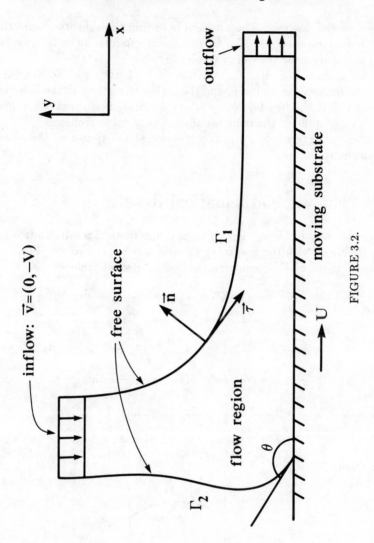

FIGURE 3.2.

region in the diagram. The problem is to find the velocity $\vec{v}$ and the two free boundary curves Γ_1, Γ_2. Of particular interest is the behavior of the free boundary near the wetting point A.

This problem was studied numerically by Kistler and Scriven [4] using domain tessellation and applying the Galerkin method in the subdomains; the fact that the free boundary is not a priori known makes the problem much more difficult than the usual boundary value problems.

Other coating models are also mentioned in [4] and in the references given there.

3.2   Known Mathematical Results

There are relatively very few rigorous mathematical results on free boundary problems with the governing equations being the Navier–Stokes equations. Solonnikov [5] considered the 2–dimensional problem

$$-\nabla^2 \vec{v} + (\vec{v} \cdot \nabla)\, \vec{v} + \nabla p = 0, \qquad \nabla \cdot \vec{v} = 0,$$

$$\vec{v} = R\, \vec{a}\,(x) \qquad \text{on} \quad \Sigma \cup \gamma,$$

$$\vec{v} \cdot \vec{n} = 0, \quad \Sigma\, \tau_i \left(\frac{\partial v_i}{\partial x_j} + \frac{\partial v_j}{\partial x_i} \right) n_j = 0 \qquad \text{on} \quad \Gamma,$$

$$\frac{d}{dx_1} \frac{\phi'(x_1)}{\sqrt{1+\phi'^2(x_1)}} - \beta\, \phi(x_1) = -p + \tfrac{1}{2} \Sigma n_i \frac{\partial v_i}{\partial x_j}\, n_j$$

where $R > 0$, $\beta > 0$ and Γ is the free boundary,

$$\Gamma\, :\, x_2 = \phi(x_1) \qquad (-1 < x_1 < 1)\, ;$$

see Figure 3.3.

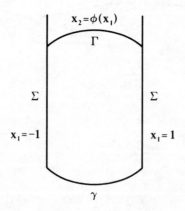

FIGURE 3.3.

The curve γ is prescribed, as well as the slopes that Γ makes with Σ; further,

$\int \vec{a} \cdot n \, ds = 0$, and

$$\int_{-1}^{1} \phi(x_1) \, dx_1 = h \quad \text{is prescribed.}$$

Solonnikov establishes the existence of a unique solution provided

$$R \quad \text{is sufficiently small} \tag{3.10}$$

A "smallness" assumption seems to appear essentially in all the mathematical treatments of free boundary problems for the Navier–Stokes equations; cf. [6], [7] [8]. Time–dependent free boundary problems for the Navier–Stokes equations were treated by Solonnikov [9], but existence was established only for small time; see also Allain [10]. We also mention the work of K. Piletskas [11], where the geometry is as described in Figure 3.4.

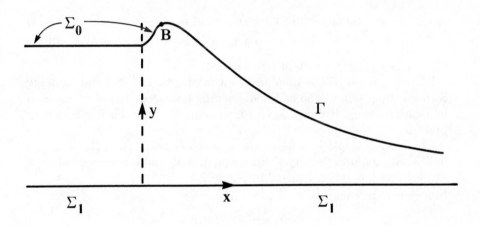

FIGURE 3.4.

On Σ_0, $\vec{v} = 0$ and on Σ_1, $\vec{v} = (\delta, 0)$ where δ is very small. The free boundary Γ starts at the point B. Although this model may well appear in a coating device, the assumption that δ is small excludes the interesting physical case.

Other results of the above types can be found in the more recent Russian literature (Solonnikov, Puhnačev, etc.). T. Beale [12] has solved the Navier–Stokes time-dependent free boundary problem in the entire R^3 space for initial data near an equilibrium.

135, 968

3.3 Simplified Models

In order to be able to do relevant mathematical analysis for coating problems it seems reasonable to consider simplified models in some flow regimes. As a first example we mention the approximation of the general steady, two–dimensional Newtonian flow by a nearly rectilinear flow

$$\delta(uu_x + vu_y) = -p_x + u_{yy} + \delta^2 u_{xx} + g\cos\phi \ , \tag{3.11}$$
$$r\delta^3(uv_x + vv_y) = -p_y + \delta^2 v_{yy} + \delta^4 v_{xx} + \delta g\sin\phi \tag{3.12}$$

where r is the Reynolds number, g is the inverse Stokes number, and the gravitational vector is $(G\cos\phi, G\sin\phi)$ (g depends on G). The fluid is nearly parallel to the x-axis with a small characteristic slope δ. On the free boundary $y = h(x)$ the following conditions hold (see [13]):

$$p + 2\delta^2 u_x \frac{1 + \delta^2 h_x^2}{1 - \delta^2 h_x^2} + \frac{\delta^3}{c}\frac{h_{xx}}{1 + \delta^2 \ h_x^2} = 0 \ , \tag{3.13}$$

$$(u_y + \delta^2 v_x)(1 - \delta^2 h_x^2) - 4\delta^2 \ h_x \ u_x = 0 \ , \tag{3.14}$$

$$v = h_x \ u \tag{3.15}$$

where c is a positive constant (the capillary number).

The system can still be simplified by assuming that $\delta^2 \ll 1$ and dropping corresponding terms. The additional assumption $r\delta \ll 1$ is often made in lubrication theory. It would be very interesting to study this free boundary problem.

The next example is a free coating problem taken from [1]; a web is pulled with uniform velocity V from a paint bath and the free boundary is divided into 3 parts as depicted in Figure 3.5.

In region 1 the equation

$$\mu \frac{d^2 u}{dy^2} - \rho g = 0$$

holds in the fluid, and the boundary conditions are

$$u = U \qquad \text{on} \quad y = 0 \qquad \text{(no slip)} \ ,$$

$$\mu \frac{du}{dy} = 0 \qquad \text{on} \quad y = H_\infty \qquad \text{(no shear exterted by the air)}.$$

We find that

$$u = U + \frac{\rho g}{\mu}\left(\frac{y^2}{\mu} - H_\infty y\right) \tag{3.16}$$

where H_∞ is yet unknown.

In region 2,

$$\mu \frac{\partial^2 u}{\partial y^2} - \rho g - \frac{\partial p}{\partial x} = 0 \ , \tag{3.17}$$

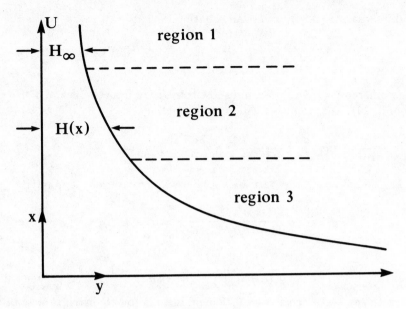

FIGURE 3.5.

$$-\frac{\partial p}{\partial y} = 0 \qquad (3.18)$$

in the fluid, and

$$u = U \quad \text{on} \quad y = 0 \ .$$

If we make the lubrication approximation mentioned above and further assume that dH/dx is very small, we get

$$T_{xy} = 0$$

and

$$p + 2\mu\,\frac{\partial u}{\partial x} = -\sigma\,\frac{d^2 H}{dx^2} \qquad (3.19)$$

on the free boundary; notice that

$$T_{yy} = -T_{xx} = -2\mu\,\frac{\partial u}{\partial x}\ .$$

We can then solve for u:

$$u = U + \frac{\rho g + dp/dx}{\mu}\left(\frac{y^2}{2} - Hy\right)\ . \qquad (3.20)$$

Introducing the *entrainment* q,

$$q = \int_0^H u\,dy\ ,$$

and integrating (3.20), we get an equation for p:

$$\frac{dp}{dx} = \frac{3\mu}{H}\left(UH - H - q - \frac{\rho g H^3}{3\mu}\right). \qquad (3.21)$$

Now differentiate (3.19) in x and substitute dp/dx from (3.21) and d^2u/dx^2 from (3.20). The result is a differential equation

$$\sigma\frac{d^3H}{dx^3} - 3\mu q\frac{d}{dx}\left(\frac{1}{H^2}\frac{dH}{dx}\right) + \frac{3\mu}{H^3}(UH - q - \frac{\rho g\ h^3}{3\mu}) = 0. \qquad (3.22)$$

In region 3 we use the static balance between gravitational and interfacial forces

$$\frac{\sigma\ d^2H/dx^2}{(1 + (dH/dx)^2)^{3/2}} = \rho g x,$$

or

$$\frac{dH/dx}{(1 + (dH/dx)^2)^{1/2}} = \frac{\rho\ g\ x^2}{2\sigma} - 1$$

since $dH/dx \longrightarrow -\infty$ as $x \longrightarrow 0$. The problem is now to match H smoothly between regions 1 and 2 (i.e., at $H = H_\infty$: $dH/dx = 0$, $d^2H/dx^2 = 0$) and between regions 2 and 3. Thus we get four boundary conditions for H, where H satisfies the third order differential equation (3.22) with unknown parameter q. But although the number of conditions is the correct one, we run into difficulties when trying to fit the parameters. Various suggestions have been made (see [1]) but the situation is still unsatisfactory. Thus it would be interesting to eliminate some of the over simplications made above in order to achieve a more convincing matching between the regions.

3.4   Future Directions

In the last 15 years there has been substantial developments in the area of free boundary problems; see the books of Kinderlehrer and Stampacchia [14] and of Friedman [15], and the references given there. Most of this work is concerned with one linear differential equation, such as $\Delta u = f$, although some nonlinear equations (such as the nonlinear elliptic equation corresponding to compressible fluid) have also been considered; see, for instance, [16]. The conditions on the free boundary are of the form

$$u = g(x), \qquad (3.23)$$
$$P(x, u, \nabla u) = 0 \qquad (3.24)$$

where g and P are given functions.

The important underlying fact is these studies is that the free boundary problem can be reformulated as a variational principle for u, namely, if u

is the minimizer of an appropriately devised functional

$$J(v) = \int F(x, v, \nabla v)$$

in an appropriate class of admissible functions then u satisfies both the differential equation and the free boundary conditions.

On the other hand free boundary problems of the form

$$P_1(x, u, \nabla u) = 0 \ ,$$

$$\tag{3.25}$$

$$P_2(x, u, \nabla u) = 0$$

(with P_1, P_2 functionally independent) cannot usually be formulated by a variational principle.

The free boundary problems for coating flow cannot be recast into a variational form. This makes their study very difficult. Thus, as a prelude to studying coating problems, one should study simpler free boundary problems which do not lend themselves to a variational principle, such as those with conditions (3.25), or with one of the conditions involving the curvature of the free boundary.

We mention here a time dependent problem for $u = u(x_1, x_2, t)$ where

$$\frac{\partial^2 u}{\partial x_1^2} + \frac{\partial^2 u}{\partial x_2^2} = 0$$

in the flow region $x_2 > h(x_1, t)$,

$$u = K(t) \qquad \text{on the free boundary } \{x_2 = h\} \tag{3.26}$$

where $K(t)$ is the curvature of the curve $x_2 = h(x_1, t)$ for any t fixed, and the velocity of the free boundary is given by $\partial u / \partial \nu$, i.e.,

$$u_{x_2} - h_{x_1} u_{x_1} = h_t \tag{3.27}$$

Local existence was proved by Duchon and Robert [17].

3.5 REFERENCES

[1] S. Middleman, *Fundamentals of Polymer Processing*, McGraw–Hill, New York, 1977.

[2] J.R.A. Pearson, *Mechanics of Polymer Processing*, Elsevier Applied Science Publishers, London, 1985.

[3] K.J. Ruschak, *Coating flows*, Ann. Rev. Fluid Mech., 17 (1985), 65–89.

[4] S.F. Kistler and L.E. Scriven, *Coating Flow theory by finite element and asymptotic analysis of the Navier–Stokes system*, Intern. J. for Numerical Methods in Fluids, 4 (1984), 207–229.

[5] V.A. Solonnikov, *Solvability of a problem on the plane motion of a heavy viscous incompressible capillary liquid partially filling a container*, Math. USSR Izvest. 14 (1980), 193–221.

[6] V.V. Puhnačev, *The plane stationary problem with free boundary for the Navier–Stokes equations*, Ž. Prikle. Mech. i Tech., # 3 (1972), 91–102.

[7] J. Bemelmans, *Liquid drops in viscous fluid under the influence of gravity and surface tension*, Manuscripta Math., 36 (1981), 105–123.

[8] J. Bemelmans and A. Friedman, *Analyticity for the Navier-Stokes equations governed by surface tension on the free boundary*, J. Diff. Eqs., 55 (1984), 135–150.

[9] V.A. Solonnikov, *Solvability of a problem on the motion of a viscous incompressible fluid bounded by a free surface*, Math. USSR Izvest., 11 (1977), 1323–1358.

[10] G. Allain, *Small–time existence for the Navier–Stokes equations with a free surface and surface tension*, in "Free Boundary Problems: Applications and Theory," vol. IV, Pitman, London, 1985, pp. 355–364.

[11] K. Piletskas, *Solvability of a 2–dimensional incompressible flow problems with noncompact free boundary*, Diff. Eqs. and their Applications, 30 (1981), 57–95.

[12] J.T. Beale, *Large time regularity of viscous surface waves*, Arch. Rat. Mech. Anal. , 84 (1983/84), 307–352.

[13] B.G. Higgins and L.E. Scriven, *Interfacial shape and evolution equations for liquid films and other viscocapillary flows*, Ind. Eng. Chem. Fundam., 18 (1979), 208–215.

[14] D. Kinderlehrer and G. Stampacchia, *An Introduction to Variational Inequalities and Their Applications*, Academic Press, New York, 1980.

[15] A. Friedman, *Variational Principles and Free Boundary Problems*, Wiley, New York, 1982.

[16] H.W. Alt, L.A. Caffarelli and A. Friedman, *Jets and cavities for compressible fluid*, J. Diff. Eqs., 56 (1985), 82–141.

[17] J. Duchon and R. Robert, *Évolution d' une interface par capillarité et diffusion de volume Existence locale en temps*, Ann. Inst. Henri Poincaré, 1 (1984), 361–378.

4

Conservation Laws in Crystal Precipitation

The material reviewed in sections 4.1–4.3 below was presented by Peter E. Castro on November 6, 1987 and is a joint work by Peter E. Castro, Anna E. Cha-Lin, David S. Ross (Applied Mathematics & Statistics) and Piotre H. Karpinski (Emulsion Engineering) at Eastman Kodak Co., Rochester, N.Y. In section 4.4 we present solutions to some of the problems raised in section 4.2.

4.1   Particles in Photographic Emulsions

A large number of industrial processes depend upon particular components for their effectiveness. In photographic emulsions the particles are microscopic grains of silver bromide, silver iodide and silver chloride all generically referred to as silver halide grains. Precipitation of these grains in controlled processes is key to high quality photographic imaging. Not only must the distribution of grain sizes be appropriate but the distribution of grain morphologies must also be controlled.

Generally, chemical engineers tend to use empirical equations to describe the evolution of the crystal size distribution. This presents difficulties in changing from one set of process conditions to another since the empirically developed equations must be re–established for the new situation. Additionally, there is little ability to predict whether new precipitation configurations will be more advantageous than the currently utilized ones. Balanced against practical use of empirical equations is the fact that full theoretical equations are generally quite complicated or even unavailable. In the drive to produce the highest quality product at the least cost it becomes ever more important to have understanding that can come from mathematical models. The models below are an introduction to the basic, almost generic, precipitation equations with ripening.

Modern photographic emulsions contain crystals with tabular morphology; see Figure 4.1, which is a photograph taken at Eastman Kodak. That is, there is a definite thickness which is much smaller than a representative dimension which might be called a diameter (much like a wafer cut from a cylinder perpendicular to the cylinder axis). Thus questions of kinetics of crystal growth with essentially two physical dimensions are of key interest to photographic scientists. What controls the growth rate of the diame-

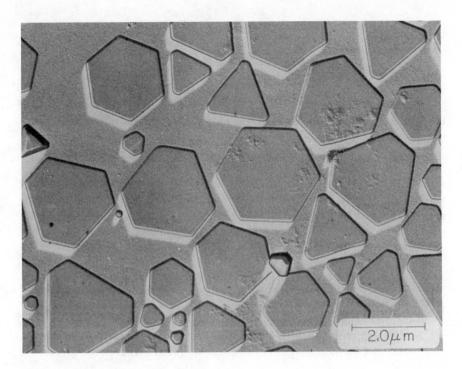

2.0 μ m

FIGURE 4.1.

ter versus thickness? At what rate do the crystals grow in general? Inverse problems of the type "given measurements of the crystal size distribution at different time points, can the growth rate parameters in the model equations be determined by a numerical optimization?" are of interest. This leads to the desire for accurate, efficient numerical algorithms to solve the precipitation equations. We begin by describing a very general setting.

Given a volume of fluid containing an amount of dissolved matter (solute) there will be in equilibrium a saturation concentration c^* which is the maximum as per unit volume of fluid that the system can hold. If the actual concentration $c(t)$ exceeds c^*, then the excess precipitates out in solid form, taken here in crystal form. Denote by $n(R,t)\, dR$ the number of crystals per unit mass of solvent which are in dR at time t ; $R = (r_1, \ldots, r_{m+3})$ is a parameter vector where three of the parameters describe the location and the remaining m parameters describe physical characteristics of the crystal.

Denote by v the phase velocity of crystals, i.e., $v = dR/dt$. Then the continuity equation for the crystal density $n(R,t)$ is

$$\frac{\partial n}{\partial t} + \text{div } (nv) - B + D = 0$$

where B and D denote the birth and death rates of crystals.

Assuming precipitation which is uniform (i.e. well–stirred solution) and taking the average of the continuity equation over the spatial variables, one obtains

$$\frac{\partial n}{\partial t} + \text{div } (n \ v_i) + n\frac{\dot{a}}{dt} \ \ell nS = B - D + F \tag{4.1}$$

where $S = S(t)$ is the total mass of the solvent in the crystal, v_i an internal variable of the crystals (size, activity, etc.), and F is the rate of injection of crystals into the system.

In order to complete the model one has to specify v_i, B, D, F.

4.2   A Simple Model of Tavare

We begin with a simple model of N.S. Tavare,

[1] Simulation of Ostwald ripening in a reactive batch crystallizer, Amer. Inst. of Chemical Engineers Journal, 33 (1987), 152–156.

It deals the phenomenon of Ostwald ripening in which small crystals tend to dissolve while large crystals grow. The crystals are balls of varying radius L (or cubes with varying side L), and

$$n(L,t) \ dL = \quad \text{no. of crystals per unit mass of solvent with}$$
$$\text{radius in} \quad (L, \ L+dL), \quad \text{at time } t.$$

Here $v_i = dL/dt$.

The driving force for the precipitation is the excess $c - c_L$ where c_L is given by the Gibbs–Thomson relation

$$c_L = c^* \ e^{\Gamma_D/L} \ ,$$

$\Gamma_D = 4 \ \sigma v/RT$, $\sigma =$ surface energy, $v =$ molecular volume of the solid, R is the gas constant and T is temperature. If $c(t) > c_L$ then material will come out of the solution and deposit onto the crystal characterized by L, and if $c(t) < c_L$ then material will dissolve from the crystal. Setting

$$L^*(t) = \frac{\Gamma_D}{\log \frac{c(t)}{c^*}} \tag{4.2}$$

and using semi-empirical growth and dissolution rates, we have

$$\frac{dL}{dt} = G(L,t) \tag{4.3}$$

where

$$G(L,t) = \begin{cases} k_g(c(t) - c^* \, e^{\Gamma_D/L})^g & \text{if} \quad L > L^*(t), \\ -k_d(c^* e^{\Gamma_D/L} - c(t))^d & \text{if} \quad L < L^*(t) \end{cases} \qquad (4.4)$$

where k_g, k_d, g, d are positive constants, and

$$1 \leq g \leq 2, \qquad 1 \leq d \leq 2 ; \qquad (4.5)$$

$d = 1$ is typical.

Equation (4.1) now becomes (after taking $\dot{S} = 0$, $B = D = F = 0$, for simplicity)

$$\frac{\partial n}{\partial t} + \frac{\partial(n\,G)}{\partial L} = 0 \qquad \text{for} \quad L > 0, \ t > 0. \qquad (4.6)$$

Finally,

$$n(L,0) = n_0(L) \geq 0 \qquad (4.7)$$

is given, and $c(t)$ is determined by

$$c(t) = c_0 + \rho k_v \int_0^\infty L^3 n_0(L) \, dL - \rho k_v \int_0^\infty L^3 n(L,t) \, dL \qquad (4.8)$$

where c_0 is the initial concentration, k_v is a geometric parameter connecting L^3 to crystal volume and ρ is the mass density of the solid phase.

Problems. (1) Prove that the system (4.2)–(4.8) has a (unique) solution.

(2) What is the asymptotic behavior of the solution as $t \longrightarrow \infty$? (It was conjectured that $n(L,t)$ approaches an atomic measure with two atoms, one at the origin and the other at some size L_∞ which needs to be determined.)

If an empirical nucleation rate

$$B = k_b(c - c^*)^b$$

(with b large) is incorporated then (4.5) is replaced by

$$\frac{\partial n}{\partial t} + \frac{\partial(n\,G)}{\partial L} = k_b(c(t) - c^*)^b \, \delta(L - L^*) \qquad (4.9)$$

where δ is the Dirac measure.

Problem (3). Solve problems (1), (2) when (4.6) is replaced by (4.9).

4.3 A More Realistic Model

In practice there is addition and removal of material taking place simultaneously with the precipitation of material onto growing crystals. Assume for simplicity three compounds (reagents) A, B, C with A, B reacting to form C, and C precipitates into crystal form. The solvent is a mixture of reagents A, B, C and a solvent which is denoted by S. The reaction kinetics of $A + B \longrightarrow C$ can be written in the form

$$\frac{dc_A}{dt} = -k_A c_A c_B + \frac{r_A}{V} + c_A \frac{\dot{V}}{V} \,,$$

$$\frac{dc_B}{dt} = -k_B c_A c_B - \frac{r_B}{V} + c_B \frac{\dot{V}}{V} \,, \qquad (4.10)$$

$$\frac{dc}{dt} = -k_C c_A c_E + \frac{r_C}{V} - c\frac{\dot{V}}{V} - P$$

where c_A, c_B, c are the concentrations of A, B, C respectively, k_A, k_B, k_C are reaction rates and r_A, r_B, r_C are rates of addition of the respective compounds from external sources. P is the change in concentration of C brought about by precipitation. Finally, the solid free-volume $V(t) = V(S(t),\ A(t),\ B(t),\ C(t))$ satisfies

$$\dot{V} = V_S \, \dot{S} + V_A \, \dot{A} + V_B \, \dot{B} + V_C \, \dot{C} \qquad (4.11)$$

where $\dot{S}$ is specified, whereas

$$\dot{A} = r_A - k_A V \, c_A \, c_B \,,$$
$$\dot{B} = r_B - k_B V \, c_A \, c_B \,, \qquad (4.12)$$
$$\dot{C} = r_C + k_C V \, c_A \, c_B - V \, P \,.$$

In order to write an expression for P, we introduce the size distribution $n(L, t)$, where $L = (L_1, \ldots, L_q)$ and, in interesting cases, $q \geq 2$. Denote by $m(L)$ the mass of the crystal characterized by L ($m(L)$ is assumed known), and set

$$\dot{m}(L) = \sum \frac{\partial m}{\partial L_j} \, G_j$$

where $(G_1, \ldots, G_m) = G = G(L, c)$ is the vector function which gives the rate of change of L (as in section 4.1). Then

$$P = \int m(L) \, B(L) \, dL + \int \dot{m}(L) \, n(L, t) \, dL. \qquad (4.13)$$

Finally, we have the differential equation for n:

$$\frac{\partial n(L, t)}{\partial t} + \operatorname{div} \left(n(L, t) \, G(L, t) \right) = B(L) - D(L) \qquad (4.14)$$

where the functions $B(L), D(L)$ are given.

Notice that if $k_C = 0, r_C = 0,\ B = D = 0$ and $V \equiv 1$ then for $q = 1$ the problem (4.10) - (4.14) reduces to the one introduced in section 4.1 (with (4.14) coinciding with (4.6)).

Problem (4). Extend problems (1), (2) to the system (4.10)–(4.14) (with $q \geq 2$).

4.4   Solution to Problems (1), (2)

We describe briefly some results obtained by Avner Friedman and Biao Oh, in response to problems (1), (2); the detailed proofs will appear elsewhere.

Theorem 4.1 *Assume that $n_0(L)$ is continuous and has compact support in*
$0 \leq L < \infty$ *and that*

$$\sup_L \ n_0(L) \ e^{d\Gamma_D/L} < \infty . \tag{4.15}$$

Then for any $d \geq 1$, $g \geq 1$ there exists a global solution $n(L, t)$ of (4.1)– (4.8); if $d > 1$, $g > 1$ then the solution is continuous.

The proof of existence for $0 \leq t \leq \eta$, η small, is obtained by a fixed point argument. For any continuous function $c(t)$ in an appropriate class K we solve (4.6),(4.7) and then correspond to the solution $n(L, t)$ the function $(Pc)(t)$ defined by

$$(Pc)(t) = c_0 + \rho k_v \int_0^\infty L^3 n_0(L) \ dL - \rho k_v \int_0^\infty L^3 n(L, t) \ dL .$$

Then one shows that P maps K into itself, P is continuous, and PK is compact. Appealing to the Schauder fixed point theorem we conclude that P has a fixed point c, i.e., $Pc = c$. The corresponding $n(L, t)$ is then the desired solution.

To prove global existence we have to extend the solution step–by–step, proving that the size of the steps remains uniformly positive in bounded time intervals. This requires some a priori estimates; in particular, we show that for any $T > 0$ there is $\delta > 0$ such that if a solution exists for $t < T$ then

$$c^* + \delta \leq c(t) \leq c_1 \qquad \text{for} \qquad 0 < t < T,$$

where

$$c_1 = c_0 + \rho k_v \int_0^\infty L^3 n_0(L) \ dL .$$

Theorem 4.2 *Assume that (4.15) and*

$$\sup_{L>0} \ L^2 \ e^{d\Gamma_D/L} \left| \frac{dn_0(L)}{dL} \right| < \infty \tag{4.16}$$

hold and that $d \geq 2$, $g \geq 2$. Then the solution of (4.1) – (4.8) is unique.

We first prove local uniqueness, by taking two solution n_1, n_2 and trying to establish an estimate of the form

$$\left| \frac{d(c_1 - c_2)(t)}{dt} \right| \leq A \sup_{0 \leq s \leq t} |c_1(s) - s_2(s)|, \qquad (4.17)$$

where $c_i(t)$ is the $c(t)$-function corresponding to n_i. The proof of (4.17) involves lengthy and somewhat delicate arguments, and relies upon the assumption (4.15); further, we can establish it only for $0 \leq t \leq \eta$, where η is a sufficiently small number. Uniqueness for $0 \leq t \leq \eta$ follows, of course, immediately from (4.16).

In order to extend the uniqueness proof for all $t > 0$, we then establish the a priori bound

$$\sup_{0 < t \leq T} \sup_{L > 0} L^2 \, e^{d\Gamma_D/L} \left| \frac{\partial n(L, t)}{\partial L} \right| < \infty$$

for any $T < \infty$.

As a by–product of the above considerations it follows that $n(L, t)$ is continuously differentiable if $g > 2$, $d > 2$.

We now turn to the asymptotic behavior as $t \to \infty$.

Theorem 4.3 *If $c(\infty) \equiv \lim_{t \to \infty} c(t)$ exists then*

$$either \qquad c(\infty) = c_1 \quad or \quad c(\infty) = c^* \ ;$$

further, if $\liminf_{t \to \infty} c(t) > c^*$ then $\lim_{t \to \infty} c(t)$ exists and equals c_1.

Theorem 4.4 *If $\lim_{t \to \infty} c(t)$ exists then*

$$\int_0^\infty n(L, t) \, dL \to 0 \quad if \quad t \to \infty$$

In proving these results we make use of the following property:

Denote by $L(\tau)$ a solution of (4.3) for $0 \leq \tau \leq t$ such that $L(\tau) > 0$ for $0 \leq \tau \leq t$. Then for any $0 < s < t$,

$$\int_{L(t)}^\infty n(y, t) \, dy = \int_{L(s)}^\infty n(y, s) \, dy \ . \qquad (4.18)$$

Avner Friedman, Biao Oh and David Ross have studied the problem (4.2) -(4.8) in the special but technologically important case where $n_0(L)$ has the form

$$n_0(L) = \sum_{m=1}^N \mu_m \, \delta(L - L_m)$$

where δ is the Dirac measure. By approximating $n_0(L)$ by smooth data they arrive at the formula

$$n(L,t) = \sum_{m=1}^{N} \mu_m \, \delta(L_m(t) - L)$$

where

$$\frac{d \, L_m(t)}{dt} = G(L_m(t), c(t)) \quad , \quad L_m(0) = L_m$$

and

$$c(t) = c_1 - \rho k_v \sum_{m=1}^{N} \mu_m \, L_m^3(t) \ .$$

They proved:

Theorem 4.5 *(i)   For any $m < N$, $L_m(t) \to 0$ in finite time; (ii)   either $x_N(t) \to 0$ in finite time or $x_N(t) \to \xi_1$ in finite time, or $x_N(t) \to \xi_2$ as $t \to \infty$, where the ξ_i are the solutions of*

$$\mu_N \, \rho k_v \, \xi^3 + c^* e^{\Gamma/\xi} = c_0 + \sum_{m=1}^{N} \rho \, k_v \, \mu_m \, L_m^3 \equiv c_1.$$

It follows that if not all the crystals decay in finite time then

$$c^* < c(\infty) < c_1 \ ,$$

a behavior which is radically different from that asserted in Theorem 4.3 in the case of $n_0(x)$ continuous.

The methods developed for proving the above results can probably be extended to apply to the differential equation (4.9).

5

A Close Encounter Problem of Random Walk in Polymer Physics

Polymers are long chain molecules with shapes as in the Figure 5.1

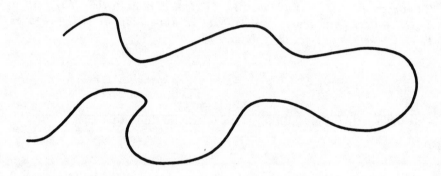

FIGURE 5.1.

The mechanical properties of polymeric materials are dominated by the entanglements of; long chain–like polymers. The entanglements impose severe restrictions on the molecular motions of polymers.

The entanglement is the basis of the unique viscoelastic behaviors observed from motion of polymer melts. Therefore, the understanding of the effects of the entanglement is the key to any molecular theory of viscoelasticity of polymers. However, the entanglement is hard to treat theoretically because polymer chains have complicated shapes.

Besides, at high density of polymers like a polymer melt, knowing configurations of a certain pair of chains is not enough because other chains are entangled with any given pair of chains. Young–Hwa Kim at a research laboratory of 3M company is attempting to find a new way of describing the entanglement problem by understanding how chains touch (close encounter) each other. He wants to obtain the probabilistic characterization of close encounters among chains and how these relax in time as a response to an external stress applied. The understanding of close encounter problem is also related to the problem of phase separation of multicomponent blends of polymers and solvents. Both problems, the prediction of viscoelastic behavior and the phase separation of polymer blends, are applicable to a wide variety of polymer research problems in industry. On November 13, 1987 Young–Hwa Kim presented a physical situation which will be described below, and posed some mathematical problems.

There is a large number of polymer chains occupying, say, the three dimensional space $\mathbb{R}^3$, and we wish to determine the probability for a pair to make n "contacts" with each other. This number of contacts is important in determining the mutual molecular interaction between them.

In order to set up the model, we divide each polymer by arcs of length b, which will will be called *segments*. The connection between two adjacent monomer in the chain has some stiffness, but if we take a side to consist of 5 or 6 monomers then the orientations between successive segments are practically uncorrelated; the corresponding length b is called *Kuhn's length*.

The next simplification is to view each side as a line segment (of length b). Thus the polymer is conceived as a polygonal curve described in Figure 5.2 with vertices $A_0, A_1, \ldots, A_N$; each vector

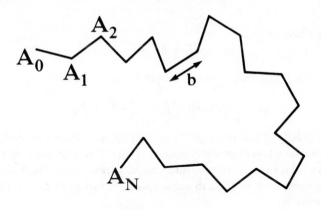

FIGURE 5.2.

$\overrightarrow{A_{i-1}A_i}$ is of length b, and

$$\text{the directions of } \overrightarrow{r}_i, \overrightarrow{r}_j \quad \text{are} \tag{5.1}$$
$$\text{stochastically independent if } i \neq j.$$

This means that the polygon

$$A \equiv (A_0, A_1, \ldots, A_N)$$

is a random walk with N steps. In the sequel we assume that all polymers in the melt consist of precisely N segments.

We can easily evaluate the length of the vector $\vec{R} = \overrightarrow{A_0 A_N}$ as follows. Writing

$$\vec{R} = \sum_{i=1}^{N} \vec{r}_i \qquad (|\vec{r}_i| = b)$$

we have

$$\vec{R} \cdot \vec{R} = \sum_{i=1}^{N} \vec{r}_i \cdot \vec{r}_i + \sum_{i \neq j} \vec{r}_i \cdot \vec{r}_j \ .$$

Taking the average $\langle \cdot \rangle$ (or the expectation) and noting that

$$\langle \vec{r}_i \cdot \vec{r}_i \rangle = \langle \vec{r}_i \rangle \cdot \langle \vec{r}_j \rangle \qquad \text{by (5.1)}$$

and

$$\langle \vec{r}_i \rangle = 0$$

provided we assume uniform distribution of the polymers, we get

$$\langle |\vec{R}|^2 \rangle = \sum_{i=1}^{N} \langle |\vec{r}_i|^2 \rangle \ ;$$

thus, on the average,

$$R = b\sqrt{N} \tag{5.2}$$

where $R = |\vec{R}|$. Similarly, for any vertices A_i, A_j we have

$$|A_i - A_j| = b|i - j|^{1/2} \qquad \text{if} \quad i \neq j \ . \tag{5.3}$$

The "excluded volume principle" implies that a polymer cannot have self-intersection (since two monomers cannot occupy the same place at the same time). If we take this restriction into consideration, then the derivation of (5.2) is incorrect and, in fact, in some sparse mixtures (5.2) is replaced by Flory's Theorem

$$R = b \, N^{0.6} \ . \tag{5.2'}$$

However in a melt with a large number of highly entangled polymers, the laws (5.2) and (5.3) are still very good approximations.

We shall denote a generic polymer in the melt by $B = (B_0, B_1, \ldots, B_N)$ and designate a particular one by $A = (A_0, A_1, \ldots, A_N)$; we shall call it a *probe polymer*. We wish to examine the number of close contacts that the polymers B make (on the average) with the probe A, when A is in "average" position.

We first define the concept of contact, which will depend on a parameter $\lambda > 0$ (for instance, $\lambda = 1$).

Suppose for some i, $|A_i - B_j| \leq \lambda b$ for k_i values $j = j_1, j = j_2, \ldots, j = j_{k_i}$. Then we say that A_i has k_i *close contacts* (or *close encounters*) with the polymer B. The number Σk_i is denoted by n, and we say that A has n close contacts with B (Then B also has n close contacts with A).

We suppose that the polymers B are uniformly distributed in all of $\mathbf{R}^3$ with density ρ; this means that in every volume dv there are approximately ρdv sides or ρdv vertices.

Problem. Find the distribution function $f(n, \rho, N)$, i.e., the average number of polymer chains which have n close contacts with an average probe polymer.

By doing numerical simulation Kim found that $f(n, \rho, N)$ is monotone decreasing in n for λ small enough. Can this be proved rigorously?

We should emphasize that in studying the quantities $f(n, \rho, N)$ we should not take A to be of special geometry; the result we seek is for an average position of the probe polymer A. Nevertheless, in order to get started, it seems reasonable to begin with the (unrealistic) simple case when the polymer chain A lies on a straight line as in Figure 5.3.

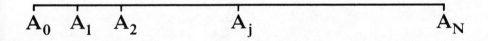

$$A_0 \quad A_1 \quad A_2 \qquad\qquad A_j \qquad\qquad\qquad\qquad A_N$$

FIGURE 5.3.

We have considered this case in subsequent discussions we had with Kim and Daniel Kleitman from M.I.T. (when he visited IMA in January and February of 1988.) Denote by p_j the probability that B makes close contact with A at the point A_j $(j \geq 1)$, and denote by f_j the probability that B makes close contact with A at the point A_j but not at any previous point A_i $(1 \leq i < j)$; we take here the situation that B_0 has close contact to A_0.

Then, under the assumption of uniform distribution of the B polymers, it was suggested by Kleitman that

$$p_j = f_j + \sum_{k<j} f_k \, p_{j-k} \tag{5.4}$$

(Strictly speaking, this formula is valid (approximately) only if j/N is small). Or, in terms of the generating functions

$$P(x) = \Sigma \, p_j x^j, \quad F(x) = \Sigma \, f_j x^i \; ,$$

$P(x) = F(x) + F(x) \, P(x)$, that is,

$$F(x) = \frac{P(x)}{1 + P(x)} \; . \tag{5.5}$$

Thus the f_j can be computed from the p_j.

To compute the p_j recall, by (5.3), that

$$|B_i - B_0| = b\sqrt{i} \; ;$$

thus B_i lies on a sphere of radius $b\sqrt{i}$. The point A_j lies a distance jb from A_0, and if B_i is to have a close contact with A_j then $\sqrt{i} \approx j$; further the area where close contact occurs is approximately the area of a cap with center A_0 and radius λb taken from the sphere of radius $\sqrt{i}$ whose area is $4\pi i$. Thus p_j is proportional to $1/i$ or

$$p_j = \frac{c}{j^2} \quad (c \quad \text{constant}) \; . \tag{5.6}$$

We can now compute the f_i from (5.5), (5.6). Finally, we introduce

$$n_j = 1 - \sum_{i=1}^{j} f_i \; ,$$

the probability that B has no contact with A_ℓ for all $1 \le \ell \le j$. Then

$1 - \sum_{i=1}^{N} f_i$ is the probability of 1 contact only (at A_0),

$\qquad \sum_{i=1}^{N} f_i n_i$ is the probability of 2 contacts,

$\sum_{0 < i_0 < j \le N} f_{i_0} f_{j-i_0} n_{j-i_0}$ is the probability of 3 contacts,

$\sum f_{i_0} f_{i_1 - i_0} f_{j-(i_1-i_0)} \, n_{j-(i_1-i_0)}$ is the probability of 4 contacts,

etc. Thus $f(n, \rho, N)$ can be determined (if n/N is small) for the case where A lies on a straight line and B_0 has close contact with A_0.

The basic problem however is still open since, when A is in general position, the computation of p_j become more complicated even if we assume that B_0 has close contact with A_0 ; also (5.4) will require justification (by an appropriate averaging argument).

An introduction to molecular theories of polymer rheology, their molecular entanglement and viscoelastic properties, is given in the book

[1] S. Middleman, The Flow of High Polymers, Interscience, Wiley, New York, 1968.

6

Mathematical Models for Manufacturable Josephson Junction Circuitry

The Josephson junction is a cryogenic device consisting of two superconductors separated by a thin gap; see Figure 6.1

The electrical resistance in the superconductors is zero and there is tunneling of electrons from one superconductor to the other. The electrons are represented by wave functions

$$\psi_j(t) = \psi_0 \, e^{i\phi_j(t)} \qquad (j = 1, 2)$$

(space-independent solutions of the free-space Schrödinger equation) in each superconductor, and there is a jump $\delta = \phi_1 - \phi_2$ in the phases of the waves across the gap, which generates supercurrent proportional to

$$e^{i(\phi_1 - \phi_2)} - e^{-i(\phi_1 - \phi_2)} \; ;$$

so the current generated is $I_0 \sin \delta$.

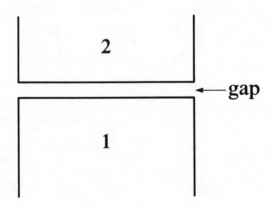

FIGURE 6.1.

Consider now a Josephson junction RCSJ (Resistance Capacitance Shunted Junction) described in Figure 6.2, consisting of capacitance C, resistance R, Josephson tunneling denoted by ×, and output voltage V and applied current I.

I is constant applied current and V is the output voltage $= \dfrac{\Phi_0}{2\pi} \dfrac{d\delta}{dt}$. The

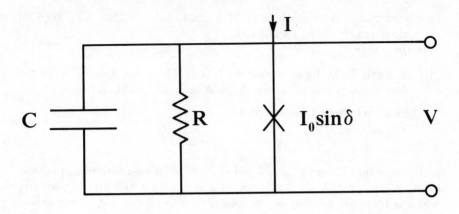

FIGURE 6.2.

currents generated in the parallel paths are

$$C \frac{dV}{dt}, \quad \frac{V}{R} \quad \text{and} \quad I_0 \sin \delta \ .$$

The voltage V is given by the formula

$$V = \frac{\Phi_0}{2\pi} \frac{d\delta}{dt} \qquad (\Phi_0 \quad \text{constant}) \ .$$

By Kirchoff's law we then get

$$I_0 \sin \delta + \frac{\Phi_0}{2\pi R} \frac{d\delta}{dt} + \frac{C \, \Phi_0}{2\pi} \frac{d^2\delta}{dt^2} = I \ . \tag{6.1}$$

For given I this is a differential equation

$$\beta_c \, \ddot{\delta} + \gamma(1+r) \, \dot{\delta} + \sin \delta = i \tag{6.2}$$

where

$$\frac{1}{R} = \frac{1}{R_s}(1+r), \quad i = \frac{I}{I_0} \ ,$$
$$\beta_c = \frac{C \, \Phi_0}{2\pi I_0}, \quad \gamma = \frac{\Phi_0}{2\pi \, I_0 \, R_s}$$

and r is a function of the voltage, in general. Setting $v = 2\pi V/\Phi_0$ we can rewrite (6.2) as a system

$$\dot{\delta} = v \ ,$$

$$\dot{v} = \frac{1}{\beta_c}[i - \sin \delta - \gamma v(1 + r(v))]. \tag{6.3}$$

One is interested in establishing stable periodic solutions for (6.2), or (6.3), for various ranges of the parameters.

In the paper

[1] M. Levi, F.C. Hoppensteadt and W.L. Miranker, Dynamics of the Josephson junction, Quart. Appl. Math., 36 (1978), 167–198.

The authors studied the equation

$$\ddot{\phi} + \sigma\dot{\phi} + \sin\phi = I$$

and proved that for any $\sigma > 0$, $I > 1$ there exists a unique running periodic solution (i.e., $\phi(t) = \omega t + p(t)$ where ω is constant and $p(t)$ is periodic) which is moreover globally asymptotically stable; if $0 < I < 1$ then this is still true if $\sigma < \sigma_0(I)$ for some $\sigma_0(I) > 0$, and no periodic solutions exists if $\sigma > \sigma_0(I)$ (For reference to earlier work see also [1]).

Applying this result to (6.2) it follows, in particular, that if $i > 1$ then unique stable running periodic solution exists if $r(v) \equiv$ const.

Question 1. Does this extend to $r(v)$ which is non-constant?

On November 20, 1987, Roger Hastings from UNISYS has presented a mathematical problem which occurs in manufacturing a system of Josephson junctions, based on SQUID (Superconductor Quantum Interference Device).

In order to describe the problem we begin by explaining what is one SQUID. Schematically it is given in Figure 6.3.

The two RCSJ have constants which are nearly the same, respectively, that is

$$C_j = C(1 + \lambda_j) \,,$$
$$\frac{1}{R_j} = \frac{1}{R}(1 + \epsilon_j) \,,$$
$$I_{0j} = I_0(1 + \xi_j) \,,$$

with $\lambda_j, \epsilon_j, \xi_j$ small, and on each side there are inductances

$$L_1 = \frac{L}{2}(1 - \eta), \quad L_2 = \frac{L}{2}(1 + \eta)$$

with η small. An external magnetic field is applied uniformly, with flux Φ_A.

As in (6.1), the currents I_j are computed by

$$I_j = I_0(1 + \xi_j)\sin\,\delta_j + \frac{\Phi_0(1 + \epsilon_j)}{2\pi R}\frac{d\delta_j}{dt} + \frac{\Phi_0 C}{2\pi}(1 + \lambda_j)\frac{d^2\delta_j}{dt^2} \qquad (6.4)$$
$$(j = 1, 2).$$

We have

$$I_1 + I_2 = I \; ; \qquad (6.5)$$

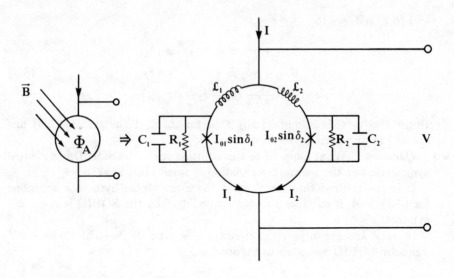

FIGURE 6.3.

also the total magnetic flux (external and induced) is zero, i.e.,

$$\frac{L}{2}(1+\eta)\,I_1 - \frac{L}{2}(1+\eta)I_2 + \frac{\Phi_0}{2\pi}(\delta_1 - \delta_2) - \Phi_A = 0 \,. \qquad (6.6)$$

If we solve for I_1, I_2 from (6.5), (6.6) and set

$$\phi_A = \frac{\Phi_A}{\Phi_0} \,, \quad \beta_c = \frac{\Phi_0\,C}{2\pi I_0} \,, \quad \gamma = \frac{\Phi_0}{2\pi\,RLI_0} \,, \quad i = \frac{I}{I_0} \,,$$

we get the system of equations

$$(1+\lambda_1)\beta_c\ddot{\delta}_1 + \gamma(1+\epsilon_1)\dot{\delta}_1 = \tfrac{i}{2} - j - (1+\xi_1)\sin\delta_1 \,,$$

$$(1+\lambda_2)\beta_c\ddot{\delta}_2 + \gamma(1+\epsilon_2)\,\dot{\delta}_2 = \tfrac{i}{2} + j - (1+\xi_2)\,\sin\delta_2$$

(6.7)

where

$$j = \frac{\Phi_0}{2\pi L\,I_0}\,(\delta_1 - \delta_2 - 2\pi\,\phi_A) - \frac{\eta i}{2} \,.$$

For ideal symmetric SQUID,

$$\lambda_j = \epsilon_j = \xi_j = \eta = 0$$

and (6.7) reduces to

$$\beta_c \ddot{\delta}_1 + \gamma \dot{\delta}_1 = \tfrac{i}{2} - j - \sin \delta_1,$$
$$\beta_c \ddot{\delta}_2 + \gamma \dot{\delta}_2 = \tfrac{i}{2} + j - \sin \delta_2, \qquad (6.8)$$
$$j = \tfrac{\Phi_0}{2\pi L I_0} (\delta_1 - \delta_2 - 2\pi \, \phi_A).$$

Roger Hastings finds numerically that for $\beta_c \ll 1$ the system (6.8) has stable running periodic solutions.

Question 2. What happens to this solution when the SQUID is not ideal symmetric but the parameter variation is small (10% say)?

It is conjectured that for $\beta_c \gg 1$ there are stable hysteretic solutions for (6.8), and, if so, does this remain valid when the SQUID is not ideal symmetric?

In some Josephson junction circuitry it is actually desirable to have $2N$ Junction SQUID as shown in Figure 6.4.

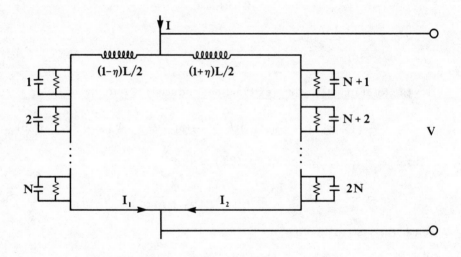

FIGURE 6.4.

Here the I_j are as in (6.4); (6.5) still holds, and (6.6) is replaced by

$$\frac{L}{2} (1 - \eta) \, I_1 - \frac{L}{2} (1 + \eta) I_2 + \frac{\Phi_0}{2\pi} \sum_{\ell=1}^{N} (\delta_\ell - \delta_{N+\ell}) - \Phi_A = 0 . \qquad (6.9)$$

Solving for I_1, I_2 we get a system of $2N$ equations

$$\beta_c(1+\lambda_m)\ddot{\delta}_m + \gamma(1+\epsilon_m)\dot{\delta}_m = \tfrac{i}{2} - j - (1+\xi_m)\sin\delta_m \; ,$$

(6.10)

$$\beta_c(1+\lambda_{N+m})\ddot{\delta}_m + \gamma(1+\epsilon_{N+m})\dot{\delta}_{N+m} = \tfrac{i}{2} + j - (1+\xi_{N+m})\sin\delta_{N+m}$$

$$(1 \le m \le N)$$

where

$$j = \frac{\Phi_0}{2\pi \, LI_0}\left[\sum_{\ell=1}^{N}(\delta_\ell - \delta_{N+\ell}) - 2\pi\phi_A\right] - \frac{\eta}{2} \; .$$

When $\lambda_\ell = \epsilon_\ell = \xi_\ell = \eta = 0$, the solution is the two   JJ SQUID with

$$L \longrightarrow NL, \quad \phi_A \longrightarrow N\phi_A \; , \quad \text{and} \quad V \to NV \; .$$

Thus, roughly speaking, the effect of the device is enhanced by a factor N. The question is what is the effect of small variations in the parameters on the efficiency of the device when the SQUID is not ideal. In particular:

Question 3. What is the bifurcation diagram of the stable periodic solution of the 2 N   JJ SQUID (with $\beta_c \ll 1$) when the variation in the parameters is small (less than 10%).

In all these problems the region around $\beta_c \sim 1$ seems (numerically) to have the most peculiar behavior.

We finally mention that systems of equations of ordinary differential equations arise in spatial discretization of the damped sine-Gordon equation

$$\phi_{tt} + \sigma\phi_t - \phi_{xx} + \sin\;\phi = 0 \qquad (0 \le x \le 1, \; t > 0)$$

which models JJ with space-dependent arguments of the wave function. Numerical work on this system appears in article [1] mentioned above. However the system is quite different from the system (6.10).

7

Image Reconstruction in Oil Refinery

7.1 The Problem

One can recover a function $F(x, y)$ defined in a domain T from the knowledge of its integrals along all lines. In fact, parametrizing the segments $\sigma_{\theta, t}$ as shown in Figure 7.1, where $\sigma_{\theta, t} = T \cap \Sigma_{\theta, t}$, the function

$$\tilde{f}(\theta, t) = \int_{\sigma_{\theta, t}} f(x, y) \, ds$$

is called the *Radon transform* of f and its inverse is given by an explicit formula (see, for instance I.M. Gelfand and G.E. Shilov, Generalized Functions, Vol. 1, Academic Press, New York, 1964).

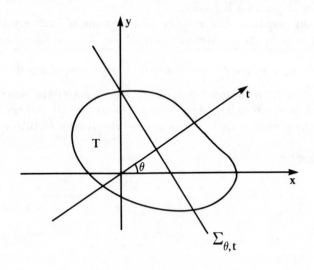

FIGURE 7.1.

In transmission tomography a source of radiation (e.g, X–rays) is directed through a sample of rays $\sigma_{\theta, t}$. The incident intensity I_0 and the transmitted intensity I_T are measured, and they are related to the linear attenuation coefficient $F(x, y)$ by

$$P(\theta, t) \equiv \log \frac{I_0}{I_T} = \int_{\sigma_{\theta, t}} F(x, y) \, ds \qquad (7.1)$$

where F is a characteristic coefficient representing internal structure of the sample ((7.1) follows from the relation $dI/ds = -FI$ with I the intensity along $\sigma_{\theta, t}$). Figure 7.2 indicates the experiment.

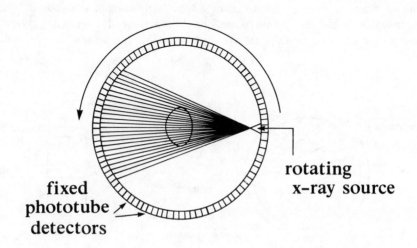

FIGURE 7.2.

As mentioned before, one can recover F from P by using inverse Radon transform. Alternatively one can recover F by using inverse Fourier transforms since the Fourier transform $\hat{F}(u, v)$ of the target F and the Fourier transform $\hat{P}(\theta, r)$ of the projection of the target $P(\theta, t)$ satisfy:

$$\hat{F}(u, v) = \iint F(x, y) e^{-2\pi i(ux + vy)} \, dx dy,$$

$$(7.2)$$

$$\hat{P}(\theta, r) = \int P(\theta, t) e^{-2\pi i r t} \, dt = \hat{F}(r \cos \theta, r \sin \theta) .$$

In practical devices however difficulties arise due to imprecise data, discrete approximations, etc. For more details we refer to the recent article:

[1] B.P. Flannery, H.W. Deckman, W.G. Roberge and K.L. D'Amico, Three-dimensional X-ray microtomography, Science, 18 Sep. 1987, vol. 237, pp. 1439–1444.

Brian Flannery from Exxon (Annadale, N.J.) has presented on December 4, 1986, a problem in oil refinery which, in contrast with the situation encountered in standard tomographic problems, is based on a very small number of samples.

In a refinery field there is typically a cylinder (not necessarily vertical) with steel wall where oil mixed with catalyst pellets is flowing. Denote the cross section by D. One would like to know the mass distribution μ of the fluid as a function of (x, y) in D. The nature of the fluid is not completely understood mathematically. It is a common practice to send beams of γ-rays along four rays; the source is seated on a collar which is mounted on the cylinder, as shown in the Figure 7.3; after measuring the transmitted intensity I_T at the end-points of the four chords, the collar is rotated by 45° and a new set of measurements are obtained, then the process is repeated, etc.

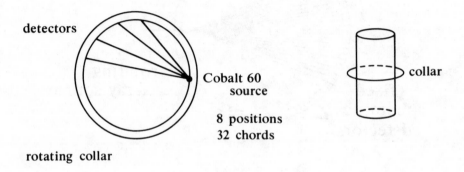

detectors

Cobalt 60
source

8 positions
32 chords

rotating collar

collar

FIGURE 7.3.

In order to measure the intensity I of the incident beams, the measurements are repeated when the cylinder is emptied. Since it is very expensive to shut off the flow of oil, one must be content to work with a small number of chords.

Denoting by σ_i $(1 \leq i \leq 32)$ the chords, the density μ satisfies (cf. (7.1))

$$\tau_i \equiv \log \frac{I}{I_T} = \int_{\sigma_i} \mu \, ds \qquad (7.3)$$

Problem. How to gain useful information on μ from the data $(\sigma_i, \ \tau_i)$.

7.2   Suggested Method

One may approximate $\mu(x, y)$ in D by polynomials $\Sigma c_{ij} x^i y^j$ and compute the coefficients from (7.3). However, this does not seem appropriate since it may well happen that μ thus computed will take negative values.

In a discussion with Brian Flannery we have come up with another approach which is described below.

Partition D into regions D_j ($1 \leq j \leq N$) for some $1 < N < 32$ and try for μ an approximation

$$\mu = A_j \quad \text{in} \quad D_j \ , \text{ i.e., } \quad \mu = \sum_{j=1}^{N} A_j \, \chi_{D_j}$$

where $\chi_{D_j} = 1$ in D_j and $= 0$ outside D_j . With the measurements τ_i defined as in (7.3), we wish to determine μ by the conditions:

$$\Phi(A) \equiv \sum_{i=1}^{32} \left[\tau_i - \int_{\sigma_i} \sum_{j=1}^{N} A_j \, \chi_{D_j} \right]^2 \quad \text{is minimized} \qquad (7.4)$$

$$\text{subject to the constraints } A_j \geq 0 \qquad \forall \, j \ .$$

Set

$$\alpha_{k\ell} = \sum_{i=1}^{32} \int_{\sigma_i} \chi_{D_k} \cdot \int_{\sigma_i} \chi_{D_\ell} \ .$$

Then

$$\sum_{k,\ell=1}^{N} \alpha_{k\ell} \, \xi_k \xi_\ell = \sum_{i=1}^{32} \int_{\sigma_i} \left(\sum_{k=1}^{N} \xi_k \, \chi_{D_k} \right)^2 \geq 0$$

and, with a careful choice of the $D_k's$ the matrix $(\alpha_{k\ell})$ is positive definite. We now define:

Step 1. Solve the quadratic programing problem (7.4), and observe that if $(\alpha_{k\ell})$ is positive definite then this is a standard problem which has been studied extensively in the literature; see, for instance,

[2] R.W. Cottle and G.B. Dantzig, Complimentary pivot theory of mathematical programming, in MAA studies in Mathematics, vol 10, Studies in Optimization, G.B. Dantzig and B.C. Eaves, editors, Amer. Math. Soc., 1974, 11, 27–51.

If a minimizing solution is such that all the A_i are positive, the $\partial\Phi/\partial A_k = 0$, i.e.,

$$\sum_{\ell=1}^{N} \alpha_{k\ell} \, A_\ell = \sum_{i=1}^{32} \tau_i \int_{\sigma_i} \chi_{D_k} \ . \qquad (7.5)$$

Denote the eigenvalues of $(\alpha_{k\ell})$ by $\lambda_1, \dots, \lambda_N$ and introduce the "accuracy number"

$$\rho = \frac{\max \lambda_i}{\min \lambda_j} \ .$$

Because of inevitable errors in the measurements, we would like to have some "stability". This is expressed by saying that ρ should not be "too large." Thus, the two quantities

$$\gamma = \min \Phi(A) \quad \text{and} \quad \rho$$

should be "as small as possible;" we may express this by means of a weight α, $0 < \alpha < 1$, setting as a goal to minimize

$$\alpha\gamma + (1 - \alpha)\rho . \tag{7.6}$$

Step 2. For each specified geometry of the chords in the collar, i.e., for each choice of angles $\theta_1, \theta_2, \theta_3, \theta_4$ which the chords make with the x–axis, compute the corresponding $\gamma = \gamma(\theta), \rho = \rho(\theta)$ (where $\theta = (\theta_1, \theta_2, \theta_3, \theta_4)$) and then find θ for which (7.6) is minimized.

This procedure should give a reasonable theorem for the optimal configuration of the collar. The corresponding components of μ are then computed as in Step 1.

In the above analysis one may also incorporate a priori partialknowledge about properties of μ. For instance, if the cylinder is in vertical position and if a region D_{i_0} lies much closer to the boundary than a region D_{j_0}, then we expect $A_{i_0} \geq A_{j_0}$ because of boundary entrainment. Thus we should incorporate the inequality $A_{i_0} \geq A_{j_0}$ into the problem (7.4).

We finally remark that the result that one may get on the optimal geometry of the chords by the above method might very well reflect the effect of the steel boundary of the cylinder on the measurements τ_i.

8

Asymptotic Methods in Semiconductor Device Modeling

For a general introduction to semiconductors see the books of Markovich [1] and Mock [2]. On January 8, 1988 Michael Ward (CALTECH and IBM at Yorktown Heights) has presented some results of his thesis [3] and of joint work by Ward, Cohen and Odeh [4] on semiconductors and posed several problems. In sections 8.1, 8.2 we shall describe his presentation for the MOSFET and PNPN problems, and in section 8.3 we shall provide a solution to the problem he posed (in section 8.1).

8.1 The MOSFET

The standard equations for semiconductor with no recombination are

$$\nabla \cdot \vec{E} = \alpha \ (p - n + N(x)), \tag{8.1}$$

$$\vec{J}_n \ = \beta(n \ \vec{E} + \eta \ \nabla n), \quad \nabla \cdot \vec{J}_n = 0 \ , \tag{8.2}$$

$$\vec{J}_p \ = -\tilde{\beta}(p \ \vec{E} - \eta \ \nabla p), \quad \nabla \cdot \vec{J}_p = 0 \ , \tag{8.3}$$

$$\vec{E} = -\nabla \psi \tag{8.4}$$

where ψ is the electric field potential, n and p are the electron and hole concentrations, N is the concentration of the implanted impurities on dopants in the channel, $\vec{J}_n$ $(\vec{J}_p)$ is the electron (hole) current density, α and η are positive physical constants, β is a function proportional to the electron mobility and $\tilde{\beta}$ is a function proportional to the hole mobility.

In n-channel MOSFET it is assumed that holes remain in equilibrium, so that $\vec{J}_p = 0$; (8.3) is then replaced by

$$p = n_i \ e^{-\psi/\eta} \quad , \quad n_i \quad \text{constant} \ . \tag{8.5}$$

The geometry of the device is given in Figure 8.1.

Equations (8.1), (8.2), (8.3), (8.5) can be written in the form of an elliptic system of two equations

$$\begin{aligned} \Delta \psi &= \alpha(n - n_i e^{-\psi/\eta} - N(x)), \\ \nabla \cdot \beta(\eta \nabla n - n \nabla \psi) &= 0. \end{aligned} \tag{8.6}$$

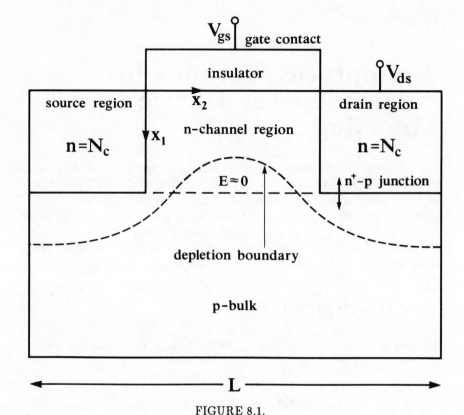

FIGURE 8.1.

We now introduce scaling which takes into account that

$$\lambda \equiv \max \left(\frac{|N(x)|}{n_i} \right) \gg 1 \qquad (\text{typically } \lambda \approx 10^6)$$

and that we have a long (or thin) channel device, so that

$$\epsilon \equiv \frac{L_D}{L} \left(\frac{\log \lambda}{\lambda} \right)^{1/2} \ll 1 \qquad \text{where} \quad L_D = \sqrt{\frac{\alpha \eta}{n_i}} \quad .$$

Set

$$\psi \; = \eta \, \log \lambda \cdot w \; , \quad n = n_i \, e^{(w - \phi_n) \log \lambda} \quad ,$$

$$x_1 \quad = L_D \left(\frac{\log \lambda}{\lambda} \right)^{1/2} x, \qquad x_2 = Ly$$

($\phi_n = $ electron quasi–Fermi potential), and

$$\frac{N(x_1, x_2)}{n_i} = -\lambda \, d(x_1) \; ;$$

the equations in (8.6) transform into

$$\widetilde{\nabla}^2 w = \frac{1}{\lambda}\left(e^{(w-\phi_n)\log\lambda} - e^{w\log\lambda}\right) + d(x), \qquad (8.7)$$

$$\widetilde{\nabla}^2 \phi_n + \widetilde{\nabla}\phi_n\cdot\left[\frac{\widetilde{\nabla}\mu_n}{\mu_n} + \log\lambda\widetilde{\nabla}(w-\phi_n)\right] = 0 \qquad (8.8)$$

where μ_n (proportional to the mobility of the electrons) is a function of x and $\partial\phi_n/\partial y$,

$$\widetilde{\nabla} = (\frac{\partial}{\partial x}, \epsilon\frac{\partial}{\partial y}),$$

and the boundary conditions are described in the Figure 8.2.

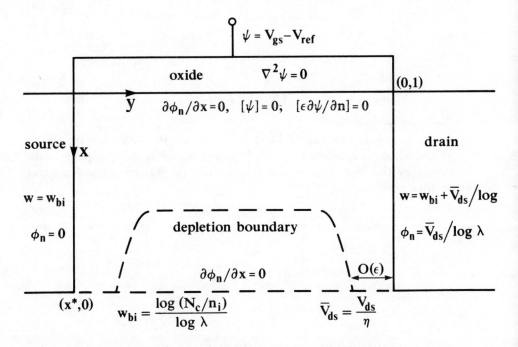

FIGURE 8.2.

Notice that (8.7), (8.8) hold for $y < 0$, whereas in the oxide region $\{y > 0\}$ w satisfies $\widetilde{\nabla}^2 w = 0$.

Michael Ward has developed asymptotic expansion for the above system, in terms of the small parameter ϵ and the large parameter λ. However, near the top corners the analysis must be modified and it leads him (after some additional scaling) to the study of the single differential equation

$$w_{xx} + w_{yy} = \frac{1}{\lambda}e^{w\log\lambda} - \frac{1}{\lambda}e^{-w\log\lambda} + 1 \qquad \text{in } R^+ = \{x > 0, \ y > 0\}, \qquad (8.9)$$

$$w = w_s \quad \text{on} \quad \partial R^+ = \{x = 0, y > 0\} \cup \{x > 0, \ y = 0\} \ , \qquad (8.10)$$

$$w(x,y) \to -1 - \sigma(\lambda) \quad \text{as} \quad x^2 + y^2 \to \infty \qquad (8.11)$$

where $\sigma(\lambda)$ is some given function of λ, $\sigma(\lambda) \sim (\lambda^2 \log \lambda)^{-1}$, so that

$$\sigma(\lambda) > 0 \ , \ \sigma(\lambda) \to 0 \quad \text{if} \quad \lambda \to \infty \ ; \qquad (8.12)$$

here w_s is a constant, $-1 < w_s < 1$. Notice that, because of (8.10), the relation (8.11) cannot hold uniformly in R^+ and must be given a more precise meaning.

Consider the free boundary problem: Find $u(x,y)$ and $f(x)$ satisfying

$$\Delta u = 1 \quad \text{if} \quad (x,y) \in R^+, \quad y < f(x),$$
$$u = -1, \ \nabla u = 0 \quad \text{on} \quad y = f(x),$$
$$u = w_s \quad \text{on} \quad \partial R^+ \ ,$$
$$f(x) = +\infty \quad \text{if} \quad 0 < x < x_0, \qquad f(x) < \infty \quad \text{if} \quad x > x_0, \qquad (8.13)$$
$$f'(x) < 0 \quad \text{if} \quad x_0 < x < \infty,$$
$$\lim_{x \to \infty} f(x) > 0$$

for some $x_0 \in (0, \infty)$.

According to Michael Ward, J.R. King and Howison (preprint) have found an explicit solution of (8.13), based on conformal mappings motivated by the work of Poluborinova–Kochina [5]. Thus the free boundary can be parametrized by

$$x(t) = \frac{2d}{\pi} K(\sqrt{1-t}), \quad y(t) = \frac{2d}{\pi} K(t) \quad (0 < t < 1)$$

where $d = \sqrt{2(1 + w_s)}$ and

$$K(z) = \int_0^{\pi/2} (1 - z \sin^2 \alpha)^{-1/2} d\alpha$$

is the complete elliptic integral of the first kind. Using

$$k(z) \sim \tfrac{\pi}{2} \left(1 + \tfrac{z}{4}\right) \quad \text{as} \quad z \longrightarrow 0,$$
$$k(z) \sim \tfrac{1}{2} \log(\tfrac{16}{1-z}) \quad \text{as} \quad z \longrightarrow 1$$

one finds that $y(t) \longrightarrow d$ as $x(t) \longrightarrow 0$; even more precisely, they get

$$\frac{y}{d} \sim 1 + \sqrt{2} \, e^{-\frac{\pi x}{2d}} \quad \text{as} \quad x \longrightarrow \infty \ .$$

Ward raised the following question:

Problem (1). Does $w(x,y) \longrightarrow u(x,y)$ as $\lambda \longrightarrow \infty$.
We shall resolve this problem affirmatively in section 8.3.

8.2   The PNPN Problem

A a single P-N Junction with recombination function and with doping
which changes sign was considered by C. Please [6] under the appropriate
scaling. The scaled equations are

$$\frac{d^2\psi}{dx^2} = n - p - \lambda N(x), \qquad 0 < x < L,$$

$$J_n = \mu_n\left(\frac{dn}{dx} - n\,\frac{d\psi}{dx}\right), \quad J_p = -\mu_p\left(\frac{dp}{dx} + p\,\frac{d\psi}{dx}\right),$$

$$\frac{dJ_n}{dx} = \frac{np-1}{\tau_n n + \tau_p p + \gamma} - d_n(\psi')|J_n| - d_p(\psi')|J_p|,$$

$$J_n + J_p = \text{const.} = J,$$

and $x_0 = 0, \ x_4 = L,$

$$N(x) = (-1)^i \quad \text{if} \quad x_{i-1} < x < x_i \qquad (i = 1, 2, 3, 4).$$

The boundary conditions are

$$np = 1, \qquad n - p - \lambda\, N(x) = 0 \quad \text{at} \quad x = 0, \ x = L,$$
$$\psi(0) = -\log\lambda,$$
$$\psi(L) = \log\lambda - V.$$

Rescaling n, p to obtain

$$\epsilon^2 \frac{d^2\overline{\psi}}{d\overline{x}^2} = \overline{n} - \overline{p} - N(x) \quad \text{on} \quad [0,1]$$

and doing the same for J_n, J_p, we get, after letting $\epsilon \to 0$,

$$\overline{n} - \overline{p} - N(x) = 0.$$

Finally, if the transition layers are replaced by jump conditions

$$[ne^{-\psi}] = [p\,e^{\psi}] = 0, \quad [J_n] = [J_p] = 0$$

then we are led to an approximate but simpler formulation of the semicon-
ductor problem in terms of a function $\sigma = n + p$ (assuming $\mu_n = \mu_p = 1, \tau_n = \tau_p = \tau$) :

$$\sigma'' - \frac{JN\,\sigma'}{\sigma^2} = 2\,R(\sigma)$$

with

$$\sigma(0) = \sigma_0(0), \ \sigma(1) = \sigma_0(1),$$
$$[\sigma^2 - N^2]_{x_i} = [\sigma' + \tfrac{NJ}{\sigma}]_{x_i} = 0 \qquad (i = 1, 2, 3)$$

where $x_0 = 0 < x_1 < x_2 < x_3 < x_4 = 1$, and

$$R(\sigma) = \frac{\sigma^2 - \sigma_0^2}{\tau(\sigma + 2/\lambda)} \ , \ \sigma_0^2 = (N^2 + \frac{4}{\lambda^2})^{1/2} \ .$$

This problem can be solved for any values of the parameters J, τ, N, λ and data $\sigma_0(0), \sigma_0(1)$, and we then form the expression

$$V = J \int\limits_0^1 \frac{d\eta}{\sigma(\eta)} + \sum_{i=1}^3 [\log(\sigma - N)] + 2 \ \log \lambda.$$

Since σ depends on J nonlinearly, the same is true of $V = V(J)$. Recall that V is the applied bias and J is the sum of the electron and hole current density. The shape of the $J - V$ curve shows how the current generated is affected by the applied bias. A basic problem is to find the shape of this curve; in particular, does there exist a branch of the $J = J(V)$ curve where $J \sim e^V$.

Problem (2). Is $V(J)$ a monotone increasing function of J? If not, is this true when we have 3 or 2 junctions (instead of 4)?

I. Rubinstein [7] has some related results.

8.3   Solution of Problem 1.

We first outline a procedure to establish existence of solutions of (8.9)–(8.11). For any $M > 0$, let

$$R_M = \{(x, y) \in R^+ \ , \qquad x^2 + y^2 < M^2\}$$

and consider (8.9), (8.10) in R_M with (8.11) replaced by

$$w = -1 - \sigma(\lambda) \quad \text{on} \quad \{x^2 + y^2 = M^2\} \ . \tag{8.14}$$

Let

$$f(w) = 1 + \frac{1}{\lambda} \ e^{w \log \lambda} - \frac{1}{\lambda} \ e^{-w \log \lambda} \ .$$

Since $f'(w) > 0$ one can establish by fixed point argument the existence of a solution $w = w_{\lambda, M}$ of the truncated problem (see [8; pp. 24–25]). Uniqueness follow by taking the difference $w_1 - w_2$ of two solutions and noting that

$$\Delta(w_1 - w_2) = f(w_1) - f(w_2) = c(w_1 - w_c)$$

where $c = c(x, y)$ is the derivative of f evaluated at some intermediate point between $w_1(x, y)$ and $w_2(x, y)$; since $c \geq 0$ the maximum principle can be applied to deduce that $w_1 - w_2 \equiv 0$.

The constant w_s is a supersolution:

$$\Delta w_s - f(w_s) = -f(w_s) = -\frac{\lambda^{w_s}}{\lambda} + \frac{1}{\lambda \lambda^{w_s}} - 1 < 0 .$$

Since $w_s \geq w$ on ∂R_M it follows, by comparison, that $w_s \geq w$ in R_M. Set $d = \inf w \equiv -1 - \gamma$. If $d < -1 - \sigma$ then the minimum of w is attained at some point $\overline{X} = (\overline{x}, \overline{y})$ inside R_M and $\Delta w(\overline{X}) \geq 0$. It follows that

$$0 \leq f(w(\overline{X})) = f(d) = 1 - \lambda^\gamma + \lambda^{-2-\gamma}$$

i.e.,

$$\lambda^\gamma - \lambda^{-2-\gamma} \leq 1 .$$

Since $\gamma > \sigma$ we conclude that

$$\lambda^\sigma - \lambda^{-2-\sigma} < 1 .$$

Assuming that $\sigma = \sigma(\lambda)$ satisfies, in addition to (8.14), the inequality

$$\lambda^\sigma - \lambda^{-2-\sigma} \geq 1 \tag{8.15}$$

(which is true of the $\sigma(\lambda)$ mentioned above, following (8.11)), we then get a contradiction, which proves that $d \geq -1 - \sigma$. Thus, if (8.14) holds then

$$-1 - \sigma \leq w \leq w_s \tag{8.16}$$

and, by monotonicity of f, also

$$f(-1 - \sigma) \leq f(w) \leq f(w_s) .$$

It follows that

$$|\Delta w_{\lambda,M}| \leq C$$

for all λ large (C constant). By L^p elliptic estimates and Sobolev's imbeding (see, for instance, [8; pp. 21–23]) we deduce that any sequence $M \to \infty$ has a subsequence such that $w_{\lambda,M} \longrightarrow w$ where w is a solution of (8.9), (8.10), and w satisfies (8.16). Since the meaning of (8.11) is not made clear, we shall not elaborate here further on the asymptotic behavior of the solution w (constructed above) at ∞; some behavior of w at ∞ can probably be obtained using appropriate comparison functions.

We now denote by w_λ *any* solution of (8.9), (8.10) (8.16) (there are infinite number of such solutions, since we do not impose a precise behavior at ∞). We shall prove

Theorem 8.1 *As $\lambda \longrightarrow \infty$, $w_\lambda \longrightarrow u$ uniformly on compact subsets of $\{x \geq 0, y \geq 0\}$, where u is the unique solution of (8.13).*

Proof. Let

$$\beta_\lambda(w) = \frac{1}{\lambda} e^{w \log \lambda} - \frac{1}{\lambda} e^{-w \log \lambda} \ .$$

Then $|\beta_\lambda(w_\lambda)| \le C$ and

$$\beta_\lambda(w) \longrightarrow \quad -\infty \quad \text{if} \quad w < -1, \quad \lambda \longrightarrow \infty,$$
$$0 \quad \text{if} \quad -1 < w < 1, \quad \lambda \longrightarrow \infty.$$

Thus $\beta_\lambda(w)$ may be viewed as a penalty term of the type used to establish existence of solutions for variational inequalities [8; pp. 24–26]. It follows that, for a sequence $\lambda \longrightarrow \infty, w_\lambda \longrightarrow u, \nabla w_\lambda \longrightarrow \nabla u$ uniformly in compact subsets of $\{x \ge 0, y \ge 0\}$ where

$$\Delta u \le 1, \quad u \ge -1,$$
$$(\Delta u - 1)(u + 1) = 0 \quad \text{a.e. in} \quad R^+$$

and $u = w_s$ on ∂R^+. By nondegeneracy [8; p. 8, Exc. 3] $u(x, y) = 0$ if dist $((x, y), \partial R^+) > N$ for some positive constant N.

Next we wish to apply the maximum principle to w_x in $\Omega = \{w > 0\}$. Observe that $\Delta w_x = 0$ in $\Omega, w_x = 0$ on $\{y = 0\}, w_x \le 0$ on $\{x = 0\}$ and $w_x = 0$ on the free boundary $\partial \Omega \cap R^+$. Further, w_x is a bounded function in Ω (by gradient estimates for solutions of variational inequalities). Since Ω is contained in a strip of width $\le N$ near ∞ (as $x \longrightarrow \infty$ or as $y \longrightarrow \infty$), by the classical Phragmén–Lindelöf theorem $w_x \longrightarrow 0$ as $x \longrightarrow \infty$ in Ω, and $\limsup w_x \le 0$ as $y \longrightarrow \infty$ in Ω. Applying the maximum principle we deduce that

$$w_x < 0 \quad \text{in} \ \Omega \ ,$$

and similarly $w_y < 0$ in Ω. It follows that there exists a function $f(x)$ such that

$$u(x, y) > -1 \quad \text{if and only if} \quad y < f(x)$$

where $f(x)$ is monotone decreasing.

Since $u = w_s$ on ∂R^+ whereas $u = -1$ on the free boundary, the free boundary has positive distance from ∂R^+; thus $f(x) = +\infty$ if and only if $0 < x < x_0$ for some $x_0 > 0$. By general results [8; Chap. 2] $f(x)$ and its inverse are analytic functions; therefore $f'(x) < 0$ for all $x > x_0$. Clearly also $f(\infty) > 0$. We finally note that

$$u(x, y) \longrightarrow w_s + \frac{1}{2} y^2 - f(\infty)y \quad \text{as} \quad x \longrightarrow \infty \quad \text{in} \ \Omega \ , \tag{8.17}$$

where

$$f^2(\infty) = 2(1 + w_s) \ . \tag{8.18}$$

Having established the existence of a solution (which necessarily must satisfy also (8.17), (8.18)), we proceed to prove uniqueness. Suppose $\overline{u}$ is another solution, $\overline{u} \not\equiv u$. Then we may assume that $u - \overline{u}$ takes positive values somewhere in R^+. Since (by (8.16), (8.17)) $u - \overline{u} \longrightarrow 0$ as $x \longrightarrow \infty$ in Ω, and similarly $u - \overline{u} \longrightarrow 0$ as $y \longrightarrow \infty$ in Ω, $u - \overline{u}$ must take positive maximum at some finite point $(\overline{x}, \overline{y})$ inside R^+. Since $u(\overline{x}, \overline{y}) > \overline{u}(\overline{x}, \overline{y}) > -1$, we have

$$\Delta u = 1 \quad \text{in a neighborhood } G \text{ of } (\overline{x}, \overline{y})$$

and thus $\Delta(u - \overline{u}) = 1 - \Delta\overline{u} \geq 0$ in G. It follows that $u - \overline{u}$ cannot take maximum at $(\overline{x}, \overline{y})$, a contradiction.

8.4 REFERENCES

[1] P. Markovich, *The Stationary Semiconductor Device Equations*, Springer-Verlag, Wien–New York 1986.

[2] M.S. Mock, *Analysis of Models of Semiconductor Devices*, Boole Press, Dublin, 1983.

[3] M. Ward, *Asymptotic methods in semiconductor device modeling*, Ph.D. dissertation, Caltech, Passadena, CA (1988).

[4] M.J. Ward, D.S. Cohen and F.M. Odeh, *Asymptotic methods for MOS-FET modelling*, preprint.

[5] Poluborinova-Kochina, *Theory of Ground Water Movement*, Princeton University Press, 1962.

[6] C. Please, *An analysis of semiconductor PN junctions*, IMA, 28 (1982), 301–318.

[7] I. Rubinstein, *Multiple steady states in one-dimensional electrodiffusion with local electroneutrality*, SIAM J. Appl. Math., 47 (1987), 1076–1093.

[8] A. Friedman, *Variational Inequalities and Free Boundary Problems*, Wiley-Interscience, New York, 1982.

9

Some Fluid Mechanics Problems in U.K. Industry

The Mathematics–in–Industry Study Group is a group of Applied Mathematicians associated with Oxford University, England who have been interacting with industry for the past 20 years, doing both modeling of industrial problems as well as solving them mathematically. One of the key figures in the group, John Ockendon, from Oxford University, gave a talk on January 15, 1988, describing several current problems. The presentation which follows is based on his talk and on his lecture notes.

9.1   Interior Flows in Cooled Turbine Blades

A turbine blade is subjected to flow of hot gas. Reducing the temperature of the hot gas, even just slightly, increases the longevity of the blade quite substantially. Methods by which to accomplish this temperature reduction are of interest to Rolls–Royce, Aero Engines, Derby; what they presently do is inject cold air from slots and holes along the surface of the blade as seen in Figure 9.1.

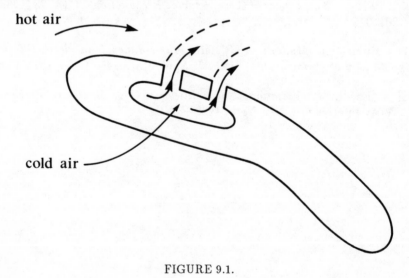

hot air

cold air

FIGURE 9.1.

The problem studied here came via Dr. T.V. Jones, Engineering Science Department, Oxford University, (see Ref. [1]) and work on it has been going on since 1983.

Consider a 2–dimensional model and the simple situation of horizontal blade surface with only one slot, as in Figure 9.2.

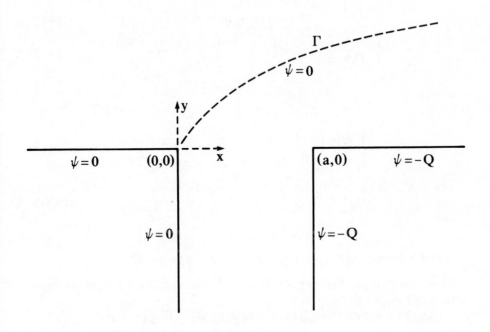

FIGURE 9.2.

The fluid is assumed to be inviscid, irrotational and incompressible, and it is further assumed that there is a sharp free boundary Γ separating the cold from the hot air, as in the diagram above. This problem was studied by Fitt, Ockendon and Jones [1]; they also justify the model in an appropriate range of parameters. In terms of the stream function ψ and the free boundary $y = f(x)$, the problem is to find $\psi(x, y)$ and f such that:

$$\Delta \psi = 0 \quad \text{in} \quad \Omega \backslash \Gamma \tag{9.1}$$

where

$$S = \{0 < x < a, \ -\infty < y < 0\} \quad \text{(the slot)},$$
$$T = \{-\infty < x < \infty, \ 0 < y < \infty\}, \quad \Omega = \ \text{int} \ (\overline{S} \cup T) \,,$$

$$\left\{ \begin{array}{l} \psi = 0 \quad \text{on} \quad \{(x, 0), -\infty < x < 0\} \cup \{(0, y); -\infty < y < 0\} \,, \\ \psi = -Q \quad \text{on} \quad \{(a, y); -\infty < y < 0\} \cup \{(x, 0); a < x < \infty\} \,, \end{array} \right. \tag{9.2}$$

$$\left\{ \begin{array}{l} \psi = 0 \quad \text{on} \quad \Gamma \,, \\ \left(\dfrac{\partial \psi^-}{\partial \nu} \right)^2 - \left(\dfrac{\partial \psi^+}{\partial \nu} \right)^2 = \lambda \quad \text{on} \quad \Gamma \,, \end{array} \right. \tag{9.3}$$

where ψ^+ is the function ψ restricted to $\{x \leq 0\} \cup \{x > 0, y > f\}, \psi^-$ is ψ restricted to $\{x > 0, y < f\}$,

$$\Gamma \quad \text{is continuously differentiable and} \quad f'(y) \geq 0 , \tag{9.4}$$

$$f(0) = 0 , \tag{9.5}$$

$$\begin{cases} f'(0) = 0 & \text{if} \quad \lambda < 0, \ f'(0) = \infty \quad \text{if} \quad \lambda > 0, \\ f'(0) = 1 & \text{if} \quad \lambda = 0 \end{cases} \tag{9.6}$$

$$h = \lim_{x \to \infty} f(x) \quad \text{is finite,} \tag{9.7}$$

$$f'(x) \longrightarrow 0 \quad \text{if} \quad x \longrightarrow \infty, \tag{9.8}$$

$$\psi(x, y) \longrightarrow \begin{cases} Q(-1 + \dfrac{y}{h}) & \text{if} \quad 0 < y < f(x), \ x \longrightarrow \infty \\ y - h & \text{if} \quad f(x) < y < C, \ x \longrightarrow \infty \ (\forall \, C > 0), \end{cases} \tag{9.9}$$

$$\nabla \psi(x, y) \longrightarrow (0, 1) \text{ if } x^2 + y^2 \longrightarrow \infty, \ \text{dist } ((x, y), \Gamma) \longrightarrow \infty, u(x, y) > 0, \tag{9.10}$$

and

$$y - h \leq \psi^+(x, y) \leq y \quad \text{in} \quad \{y > 0\} . \tag{9.11}$$

The following results were established by Friedman [2]:

Theorem 9.1 *(i) For any* $-1 < \lambda < \infty$ *there exists a unique solution* (Q, ψ, f) *of (9.1)–(9.11).*
(ii) The function $Q = Q(\lambda)$ *is monotone increasing in* λ *and*

$$Q(\lambda) \longrightarrow 0 \quad \text{if} \quad \lambda \longrightarrow -1$$

$$\frac{Q(\lambda)}{\sqrt{\lambda}} \longrightarrow \gamma \quad \text{if} \quad \lambda \to \infty$$

where γ *is a positive constant determined by solving a limiting free boundary problem.*

Of particular interest is the asymptotic behavior of $Q = Q(\lambda)$ and of the limiting height $h = h(\lambda)$ of the free boundary as $\lambda \longrightarrow -1$.

Theorem 9.2 *[2] There exist positive constants* c, C *such that*

$$c(1 + \lambda)^{3/2} \leq Q(\lambda) \leq C(1 + \lambda)^{3/2} , \tag{9.12}$$

$$c(1 + \lambda) \leq h(\lambda) \leq C(1 + \lambda) , \tag{9.13}$$

as $\lambda \longrightarrow -1$; *consequently*

$$h(\lambda) \asymp (Q(\lambda))^{2/3} \quad \text{as} \quad Q(\lambda) \to 0 . \tag{9.14}$$

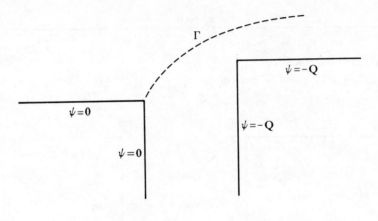

FIGURE 9.3.

The second inequality in (9.12) was suggested by formal considerations in [1].

Problem (1). Extend the results of Theorems 9.1, 9.2 to more general geometries, such as in Figure 9.3.

It was proved in [2] that for any $0 < Q < \infty$ there exists a solution (λ, ψ, f) of the problem (9.1)–(9.11) when the surface of the blade $\{y = 0\}$ is replaced by any smooth curve $y = k(x)$ provided $k(+\infty)$ and $k(-\infty)$ exist. Thus what needs to be proved for the above geometry is uniqueness and then extensions of Theorem 1 (ii) and Theorem 9.2.

Recently Dewynne, Howison and Ockendon [3] considered a related suction problem also arising in turbine blade cooling, where cold air is sucked into a slot as shown in Figure 9.4
with $\psi = 0$, $\quad |\nabla\psi| = c$ on the free boundary Γ $\quad (c > 1)$; $\psi \equiv 0$ below the free boundary. The quantity $\epsilon^2 \equiv c^2 - 1$ measures the strength of the suction. The stream line $\{\psi = m\}$ intersects the fixed boundary at a point B which depends only on c (assuming m is given); the stagnation point B may be either below A or to the right of A. They study the dependence of B on c and show, in particular, that B does not vary monotonically with c.

Problem (2). Extend the results of [3] to the geometry of Figure 9.4

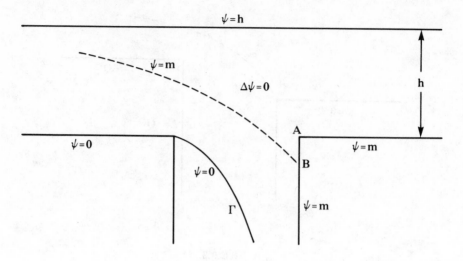

FIGURE 9.4.

9.2 Fiber Optic Tapering

The problem was introduced by Dr. W. Stewart of the Plessey Co., Towces-
ter, Northants in 1987 A nearly cylindrical fiber is made by drawing from
a container of molten glass; see Ref. [4] [5].

Then the fiber is thinned by pulling with electric motors at $x = s_i(t)$.
The subsequent cross section may be as shown in Figure 9.5, which has
undesirable optical properties resulting from uneveness.

A very simple model is to assume undirectional flow $u(x)$ and an approximate conservation of mass and axial momentum:

$$A_t + (uA)_x = 0 , \tag{9.15}$$

$$(3\mu A u_x)_x = 0 \tag{9.16}$$

where 3μ is the so-called Trouton viscosity. We also prescribe initial and
boundary data,

$$A(x,0) = A_0 (x) , \quad u = \dot{s}_i(t) \quad \text{at} \quad x = s_i(t) \quad (i = 1, 2) \tag{9.17}$$

and assume symmetry with respect to $x = 0$, so that $u(0, t) = 0$.

The case $\mu = $ const. was considered by Dewynne, Ockendon and Wilmott
[6]. We can set

$$A u_x = a(t) \quad (a(t) \quad \text{unknown as yet}),$$

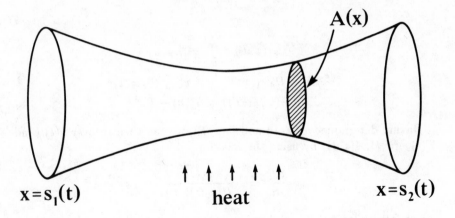

FIGURE 9.5.

$$T = \int_0^t a(\tau) \, d\tau \,, \tag{9.18}$$

$$u = a(t)U$$

and then the system becomes

$$\frac{dA}{dT} \equiv \left(\frac{\partial}{\partial T} + U \frac{\partial}{\partial x} \right) A = -1 \,, \tag{9.19}$$

$$A \, U_x = 1 \,, \tag{9.20}$$

$$A(x,0) = A_0(x), \quad U(0,T) = 0 \,. \tag{9.21}$$

If $x(T)$ is the solution of

$$\frac{dx}{dT} = U(x,T) \,, \quad x(0) = x$$

then (9.19) becomes $dA(x(T),T)/dT = -1$, so that

$$A(x(T),T) = A_0(x) - T \,, \tag{9.22}$$

and (9.20) yields

$$U(x,T) = \int_0^x \frac{d\xi}{A(\xi,T)} \,. \tag{9.23}$$

We can now solve (9.19)–(9.21) for all $-\infty < x < \infty$ and T small (such that $T < \inf A_0(x)$) by successive iterations:

$$U_{n+1}(x,T) = \int\limits_{0}^{x} \frac{d\xi}{A_n(\xi,T)} \ ,$$

$$\frac{dx_{n+1}}{dT} = U_{n+1}(x,T), \qquad x_{n+1}\,(0) = x,$$

$$A_{n+1}(x_{n+1}(T)) = A_0(x) - T \ .$$

Having determined $A(x,T)$ and $U(x,T)$, we can then recover $a(t)$ (and $T(t), u(x,t), A(x,t)$) by using the relation

$$\frac{dT}{dt} = \frac{\dot{s}_2(t)}{U(s_2(t),\ T)}$$

which follows from (9.18) and $u_x(s_2(t),t) = \dot{s}_2(t)$. As shown in [6], working with the independent variables A,T one can actually very quickly solve (9.19), (9.20) in the form

$$A_x = AF(A+T) \tag{9.24}$$

where F, an arbitrary function, is to be determined by the initial condition $A(x,0) = A_0(x)$. Using (9.24) one can effectively study properties of the solution; for instance if $A_0(x) = 1 + x^\alpha$ $(\alpha \geq 1)$ then $A \to 0$ implies $s \to \infty$. (See [6] for other results).

Problem (3). Study the behavior of solutions of (9.15)–(9.17) in case $\mu = \mu(x,t)$.

J. Keller (Stanford University) and L. Ting (N.Y.U) are studying the effects of surface tension on the tapering problem.

9.3   Ship Slamming

A problem of interest to the Royal Navy is ship slamming. The study was proposed by Mr. D. Chalmers, Ministry of Defence, Foxhill, Baltr. in 1987. Already in the case of a very simple model of a wedge–shaped hull (see Figure 9.6), this problem is very difficult; it is known as the *wedge-entry problem*, and was considered as early as 1932; see [7]. The angle $\frac{\pi}{2} - \alpha$ is called *deadrise* angle and may be small. The velocity potential ϕ is harmonic and satisfies

$$\begin{cases} \dfrac{\partial f}{\partial t} + \nabla\phi \cdot \nabla f = 0 \ , \\[2mm] \dfrac{\partial \phi}{\partial t} + \dfrac{1}{2}\,|\nabla\phi|^2 + gy = 0 \\[1mm] \qquad\qquad \text{on the free boundary} \qquad f(x,y,t) = 0 \ . \end{cases} \tag{9.25}$$

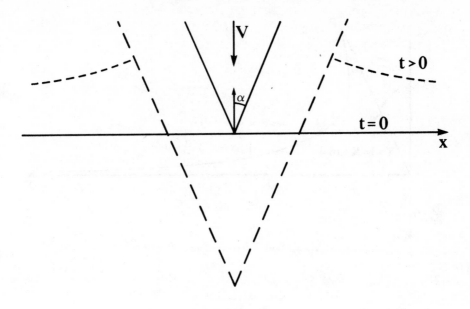

FIGURE 9.6.

If the slamming velocity V is large then the gravity term gy may be dropped. In this case it is well known [8] [9] that one can transform the problem into a 2–dimensional problem by introducing similarity variables $X = x/Vt$, $Y = y/Vt$. If we set

$$\phi = V^2 t \Phi(X, Y)$$

and if we write

$$\Phi = -\frac{1}{2}(X^2 + Y^2) + \chi ,$$

then the free boundary problem for χ in the (X, Y)-plane is described in Figure 9.7, with $\Delta\chi = -2$ in the fluid, $\chi \sim -\frac{1}{2}(X^2 + Y^2)$ at ∞, and $\frac{\partial\chi}{\partial\nu} = 0$, $\left(\frac{\partial\chi}{\partial s}\right)^2 + 2\chi = 0$ on the free boundary Γ, where s measures the arc–length (the last relation implies $\chi = -\frac{1}{2}s^2$).

The angle β at the initial point of the free boundary is not a priori known, but $\alpha + \beta < \frac{\pi}{2}$ (Mackie [8]).

For *harmonic* function χ, free boundary conditions of the form

$$\frac{\partial\chi}{\partial\nu} = 0 , \quad \left(\frac{\partial\chi}{\partial s}\right)^2 = Q$$

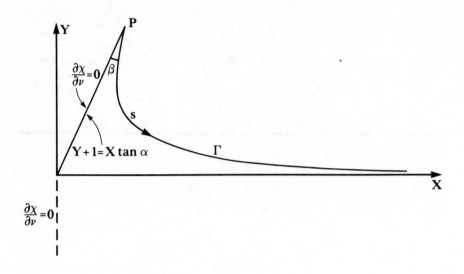

FIGURE 9.7.

can be transformed into

$$\phi = 0 \, , \qquad \left(\frac{\partial \phi}{\partial \nu}\right)^2 = Q \qquad (9.26)$$

for the harmonic conjugate ϕ of χ, and the free boundary problem for ϕ can be reformulated in a variational form (see [10], [11; Chap. 3]). If we change the equation for χ above into $\Delta \chi = 0$, then we can reduce the problem into a simpler problem with free boundary conditions (9.26) where $Q = -2\chi$ is a functional of ϕ. A study of this problem may possibly be helpful for the study of the wedge–entry problem.

9.4 REFERENCES

[1] A.D. Fitt, J.R. Ockendon and T.V. Jones, *Aerodynamics of slot film cooling Theory and experiment*, J. Fluid Mech., 160 (1985), 15–27.

[2] A. Friedman, *Injection of ideal fluid from a slot into a free stream*, Archive Rat. Mech. Anal., 94 (1986), 335–361.

[3] J.N. Dewynne, S.D. Howison and J.R. Ockendon, *Slot suction from inviscid channel flow*, J. Fluid Mech., to appear.

[4] J.R.A. Pearson and M.A. Matovich, *Spinning a molten thread line*, Ind. Engin. Chem. Fund., 8 (1969), 605–609.

[5] F.T. Geyling and G.M. Homsy, *Extensional instabilities of the glass fibre drawing process*, Glass Tech., 21 (1980), 95–102.

[6] J. Dewynne, J.R. Ockendon and P. Wilmott, *On a mathematical model for fiber tapering*, to appear.

[7] H. Wagner, *Über Stoss–und Gietvorgänge an der Oberfäcke von Flüssigkleiten*, Z. Angew. Math. Mech., 12 (1932), 193–215.

[8] A.G. Mackie, *The water entry problem*, Quart. J. Mech. Appl. Math. 22 (1969), 1–17.

[9] A.B. Taylor, *Singularities at flow separation points*, Quart. J. Mech. Appl. Math., 26 (1973).

[10] H.W. Alt and L.A. Caffarelli, *Existence and regularity for a minimum problem with free boundary*, J. Reine Angew. Math., 325 (1981), 105–144.

[11] A. Friedman, *Variational Principles and Free–Boundary Problems*, Wiley, New York, 1982.

10

High Resolution Sonar Waveform Synthesis

The problem described in this chapter was presented by Craig Poling from Honeywell on January 29, 1988.

A narrow band signal is one with frequency f_c, where $f_c \gg B$, B the frequency band. The signal is represented in the form

$$s(t) = a(t) \cos 2\pi f_c t + b(t) \sin 2\pi f_c t$$
$$= Re\{\tilde{u}(t)e^{2\pi i f_c t}\} = Re\{\psi(t)\}$$

where $\tilde{u}(t) = a(t) - ib(t)$. The goal is to detect the character of an object (for instance, a mobile underwater object) by sonar waveform synthesis, as shown in Figure 10.1

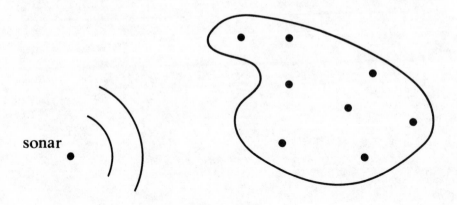

FIGURE 10.1.

In order to achieve this goal one collects data from a number of reflectors within the object. The return signal from the j-th reflector is

$$s_{\tau_j, \Delta f_j}(t) = Re\{A(\tau_j, \Delta f_j)\tilde{u}(t - \tau_j)e^{2\pi i f_j(t - \tau_j)}\}$$

where $f_j = f_c - \Delta f_j$, $\Delta f_j \approx -2v_j f_c/c$ (v_j = relative velocity of the reflector, c = speed of sound), $\tau_j = 2r_j/c$ (r_j = distance from the sonar source to the reflector).

Setting
$$D(\tau_j, \ \Delta f_j) = A(\tau_j, \Delta f_j)e^{2\pi i f_j \tau_j}$$

we get
$$s_{\tau_j, \Delta f_j}(t) = Re\{D(\tau_j, \Delta f_j)\ \psi(t - \tau_j)e^{-2\pi i \Delta f_j t}\}\ . \tag{10.1}$$

The return from the object, $s_0(t)$, is defined by

$$s_0(t) = \sum_j s_{\tau_j, \Delta f_i}(t)\ . \tag{10.2}$$

In many situations we have some knowledge of $\tau_j, \Delta f_j$, and we seek to determine the amplitude of the reflection from individual reflectors, $|D(\tau_j, \Delta f_j)|$; this is, for instance, the case in tomography. From the measurements of $s_0(t)$ we construct the integral

$$I(\tau, \Delta f) = \int\limits_{-\infty}^{\infty} s_0(t)\ \psi^*(t - \tau)e^{2\pi i \Delta f \cdot t}$$

$$= \sum_j D(\tau_j, \Delta f_j) \int\limits_{-\infty}^{\infty} \psi(t - \tau_j)\psi^*(t - \tau)e^{2\pi i(\Delta f - \Delta f_j)t}\ dt$$

where ψ^* is the complex conjuage of ψ. A change of variables $t = s + \frac{1}{2}(\tau_j + \tau)$ easily yields

$$I(\tau, \Delta f) = \sum_j D(\tau_j, \Delta f_j)e^{\pi i(\Delta f - \Delta f_j)(\tau_j + \tau)}\theta_{\tilde{u}}(\tau_j - \tau, \Delta f_j - \Delta f) \tag{10.3}$$

where

$$\theta_{\tilde{u}}(\alpha, \beta) = \int\limits_{-\infty}^{\infty} \tilde{u}(s - \frac{\alpha}{2})\tilde{u}^*(s + \frac{\alpha}{2})e^{-2\pi i \beta s}\ ds \tag{10.4}$$

is the so-called *ambiguity function* of the sonar waveform. The function $(\tau, \Delta f) \to I(\tau, \Delta f)$ is called the *image of the object in the range–doppler space*, and the problem is how to device $\tilde{u}$ (which is at our disposal) so as to be able to resolve the amplitudes $|D(\tau_k, \Delta f_k)|$ from the image $I(\tau, \Delta f)$.

If we could choose $\tilde{u}(t)$ in such a way that

$$\theta_{\tilde{u}}(\alpha, \beta) = \begin{cases} 1 & \text{if } \alpha = \beta = 0 \\ 0 & \text{elsewhere ,} \end{cases} \tag{10.5}$$

then $|D(\tau_k, \Delta f_k)|$ can be resolved from

$$|I(\tau_k, \Delta f_k)| = |D(\tau_k, \ \Delta f_k)|\ .$$

However (10.5) can never be achieved, and instead we wish to achieve, with some control $\tilde{u}$, an ambiguity function $F(\tau, \omega)$ which is a bell-shaped approximation to (10.5)

Let K denote a class of controls $\tilde{u}$ which are feasible technologically. K may include functions of the form

$$\tilde{u}(t) = \begin{cases} 1 & \text{if } t < \tau \\ 0 & \text{if } t > \tau, \end{cases}$$

$$\tilde{u}(t) = e^{2\pi i \mu t^2},$$

bell–shaped curves with compact support, etc. The problem then is to minimize the functional

$$J_p(\tilde{u}) \equiv \left\{ \int_{-\infty}^{\infty} \int_{-\infty}^{\infty} |F(\tau,\omega) - \theta_{\tilde{u}}(\tau,\omega)|^p \, d\tau d\omega \right\}^{1/p} \tag{10.6}$$

or

$$J_\infty(\tilde{u}) = \sup_{-\infty < \tau, \omega < \infty} |F(\tau,\omega) - \theta_{\tilde{u}}(\tau,\omega)|, \tag{10.7}$$

when $\tilde{u}$ varies in K.

For the case $p = 2$ Sussman [1] has studied the problem by decomposing $\tilde{u}(t) = \Sigma \, \alpha_j \phi_j(t)$ by means of an orthonormal base in $L^2(-\infty, \infty)$. Then

$$\theta_{\tilde{u}}(\tau,\omega) = \Sigma \, \alpha_j \, \alpha_\ell \, K_{j\ell}(\tau,\omega)$$

where

$$K_{j\ell}(\tau,\omega) = \int_{-\infty}^{\infty} \phi_j(t - \frac{\tau}{2})\phi_\ell^*(t + \frac{\tau}{2})e^{-2\pi i \omega t} \, dt.$$

This method was tried numerically by Craig Poling for the desired ambiguity function "close" to (10.5), and it was moderately successful, and, in fact, superior to other commonly used methods.

However L^2 and, in fact, any L^p norm does not eliminate side–lobes and, as a consequence, strong scatterers mask nearby weak scatterers. Thus, one is really forced into studying the following:

Problem. Find $\tilde{u}$ in K which minimizes $J_\infty(\tilde{u})$.

Existence of a solution can be established by taking a minimizing sequence and extracting a convergence subsequence. The real interest here is in proving uniqueness of a minimizer, in deriving useful necessary conditions for a minimizer, and in devicing numerical algorithms to compute it.

Since the L^∞-norm is not differentiable, one cannot derive the usual optimality conditions by small perturbation of the minimizer. To overcome this difficulty one may perhaps begin by studying the L^p minimization problem, derive some necessary conditions, and then carefully analyze what happens as $p \to \infty$.

But it is perhaps also possible to deal directly with the L^∞ minimization problem. In the case of one–dimensional variable, using Tschebyscheff's polynomials one can solve L^∞ minimization problems for functionals

$$\tilde{J}(\tilde{u}) \equiv \sup |F - \tilde{u}|$$

when F is given and $\tilde{u}$ varies in some "nonlinear class"; see the book of Davis [2; section 7.7] and the references given in [2; p. 155].

10.1 REFERENCES

[1] S.M. Sussman, *Least–square synthesis of radar ambiguity functions*, IRE Transactions on Information Theory, vol. IT–8 (1962), 246–254.

[2] P.J. Davis, *Interpolation and Approximation*, Dover Publications, New York, 1975.

11

Synergy in Parallel Algorithms

In parallel computations one wishes to devise algorithms which will introduce some cooperation among the various processes. Although this will increase the complexity, it might improve the error reduction to such an extent so as to render the total performance beneficial. In such a case we speak of "synergistic algorithm" (synergism is "a cooperative action so that the total effect is greater than the effects taken independently"). This concept with some examples and applications was introduced by Henderson and Miranker [1]. On February 5, 1988 Willard Miranker from IBM Thomas J. Watson Research Center (Yorktown Heights) surveyed this work, part of which we describe below, and outlined some open problems.

11.1   General framework

Consider iterative algorithm

$$x^{k+1} = f_k(\theta_k, x^k)$$

where θ_k is chosen at the k-th step so as to produce from x^k the next approximation x^{k+1}; an error function $\phi(x)$ is given. After M steps we obtain

$$x^k \to x^{k+1} \to \ldots \to x^{k+M} .$$

Alternatively we can use a composite transformation with M processors which produces x^{k+M} from x^k, say

$$x^{k+M} = F_{k,M} \left(\underline{\theta}_k, x^k \right) \qquad (\underline{\theta}_k \quad \text{is a vector}).$$

We now want to use a composite synergistic transformation $x^k \to x_s^{k+M}$ and compare it with the x^{k+M} obtained above either by the serial or parallel algorithm. The comparison is done by means of the quantity

$$s = \operatorname*{average}_{k} \frac{\phi(x^{k+M}) - \phi(x_s^{k+M})}{\phi(x_s^{k+M})} \tag{11.1}$$

which is called *synergy*.

We define the convergence rate of a composite of M steps by

$$C = -\operatorname*{average}_{i} \log \frac{\phi(x^{i+M})}{\phi(x^i)} .$$

That is, if
$$\phi(x^{i+M}) = e^{-C_i}\,\phi(x_i)$$

then C = average C_i. After N composite steps the error is reduced on the average according to

$$\phi(x^{i+NM}) = e^{-CN}\phi(x_i) = \delta\phi(x^i) \tag{11.2}$$

where
$$N = -\frac{\log\delta}{C}\quad.$$

The total time required to perform these N steps in conventional parallel mode is

$$T = NW_M$$

where W_M is the work (measured by time) it takes to perform one composite step.

Similarly we define the quantities C_s, N_s, W_M^s, T_s for the synergized algorithm corresponding to the same error reduction factor δ. The relative complexity (synergized/conventional) is

$$\frac{T_s}{T} = \frac{N_s}{N}\frac{W_M^s}{W_M} = \frac{C}{C_s}\frac{\sigma}{\sigma_s} \tag{11.3}$$

where
$$\sigma = \frac{W}{MW_M}\quad,\quad \sigma_s = \frac{W}{MW_M^s} \tag{11.4}$$

measure, respectively, the efficiency of the conventional parallel algorithm and the synergized algorithm relative to the serial algorithm; here W is the work (or complexity) of the serial algorithm in M steps. If we wish to compare a serial algorithm with synergistic parallel algorithm, then we simply take $T = NW$, and

$$\sigma = 1,\quad \sigma_s = \frac{W}{MW_M^s}\quad. \tag{11.5}$$

From (11.1),(11.2) we have

$$C_s = C + \log(1+s)$$

and using this in (11.3) we get

$$\frac{T_s}{T} = \frac{\sigma/\sigma_s}{1 + \frac{\log(1+s)}{C}}\quad,$$

or
$$s = e^{C(\sigma T - \sigma_s T_s)/\sigma_s T_s} - 1 \tag{11.6}$$

where σ, σ_s are defined either by (11.4) or by (11.5). The synergistic algorithm is profitable if $T_s \leq T$ (recall that the error reduction δ is fixed for both algorithms). Thus the breakeven level of synergy $(T = T_s)$ is

$$s^* = e^{C(\sigma - \sigma_s)/\sigma_s} - 1 \qquad (11.7)$$

We summarize: A synergistic algorithm is ahead of the conventional algorithm if and only if $s \geq s^*$.

11.2   Gauss–Seidel

Consider a system of N linear equations

$$Ax = b \qquad (11.8)$$

and choose a basis $\{\hat{e}_i\}$ in $\mathbf{R}^N$. In the Gauss–Seidel algorithm the sequence of iterates is

$$x^k = x^{k-1} + \theta_k \, \hat{e}_k$$

where θ_k is chosen so that

$$\hat{e}_k^*(A(x^{k-1} + \theta_k \hat{e}_k) - b) = 0 \qquad (k = 1, 2, \dots) \ .$$

Thus

$$\theta_k = \frac{\hat{e}_k^*(b - Ax^{k-1})}{\hat{e}_k^* A\hat{e}_k} = \frac{1}{a_{kk}} \left(b_k - \sum_{j=1}^{N} a_{kj} \, x_j^{k-1} \right) .$$

Set $E^k = x^k - x^0$, where x^0 is the solution of (11.8). Then $\|E^k\|$ is the error $\phi(x^k)$. We compute

$$E^k = E^{k-1} - \frac{\hat{e}_k^* A \, E^{k-1}}{\hat{e}_k^* A e_k} \, \hat{e}_k \ .$$

Taking the norm one obtains (see [1])

$$\|E^k\|^2 = \|E^{k-1}\|^2 - (\hat{e}_k^* E^{k-1})^2 + O(\epsilon)\|E^{k-1}\|^2$$

where

$$\epsilon = \sup_k \frac{\|\hat{e}_k^k A(I - \hat{e}_k \hat{e}^*)\|}{\hat{e}_k^* A\hat{e}_k}$$

measures the diagonal dominance of A. After a complete sweep (i.e., N steps) the error is reduced by a factor which is $O(\epsilon)$.

We next devise a synergistic algorithm which is a block Gauss–Seidel version. We partition the base $\{\hat{e}_i\}$ into disjoint sets and denote by I_k the matrix whose columns are the basis vectors from the k-th set. The k-th step is

$$x^k = x^{k-1} + I_k \underline{\theta}_k$$

where $\underline{\theta}_k$ is chosen so that

$$I_k^*(A(x^{k-1} + I_k \, \underline{\theta}_k) - b) = 0 \; .$$

By similar computations as above we find that after a complete sweep the error is reduced by a factor which is $O(\bar{\epsilon})$, where

$$\bar{\epsilon} = \sup_k \; \|(I_k^* \, AI_k)^{-1} I_k^* \, A(I - I_k I_k^*)\|$$

measures the block diagonal dominance of A. Thus, by (11.1),

$$s = \frac{O(\epsilon)}{O(\bar{\epsilon})} - 1$$

and one may reasonably guess that for "most" matrices $s = O(\epsilon/\bar{\epsilon}) - 1$. Banded matrices are matrices (a_{ik}) with

$$a_{i,i+j} = d_j \; , \qquad a_{\ell,\ell-j} = d_j \; ,$$
$$d_0 \neq 0, \qquad d_1 \neq 0, \ldots, \qquad d_k \neq 0, \; d_i = 0 \quad \text{if} \quad i > k \; .$$

Computations carried out in [1] comparing the serial Gauss–Seidel algorithm with the synergistic block Gauss–Seidel algorithm for random band matrices, show that the synergy increases as the size of block increases, and since communication costs are neglected, $s > s^*$,

11.3   The Heat Equation

Consider the initial value problem

$$u_t = u_{xx} \quad \text{for} \quad -\infty < x < \infty, \; t > 0,$$

$$\text{(11.9)}$$

$$u(x,0) = \psi(x) \quad \text{for} \quad -\infty < x < \infty \; .$$

A finite difference scheme of the form

$$U(x, t + \Delta t) = \sum_{j=-J}^{J} a_j U(x + j\Delta x, t) \; ,$$
$$U(x,0) = \psi(x)$$

is called a *two-level explicit scheme*; the error is taken as some norm of

$$E(x,t) = u(x,t) - U(x,t) \; .$$

There are various ways to determine the coefficients a_j. The conventional method is to consider

$$E(x, \Delta t) \equiv u(x, \Delta t) - \sum_{j=-J}^{J} a_j \psi(x + j\Delta x),$$

expand by Taylor's theorem, using $u_t = u_{xx}$, and equate to zero a number $q \leq 2J + 1$ of the low order powers of $\Delta x, \Delta t$. This approach is limited by the size of the high derivatives of u and thus, in particular, will not work well for fast oscillating data ψ. In order to device a good method for such $\psi's$, we use L^2 estimates, setting

$$e(t) = \{ \int_{-\infty}^{\infty} |u(x,t) - U(x,t)|^2 \, dx \}^{1/2} .$$

Then, the a_j which minimize $e(\Delta t)$ are determined by

$$\sum_{j=-J}^{J} a_j \, b_{k-j} = c_k \qquad (-J \leq k \leq J) \tag{11.10}$$

where

$$b_k = b_{-k} = \int_{-\infty}^{\infty} \psi(x)\psi(x + k\Delta x) \, dx , \tag{11.11}$$

$$c_k = \int_{-\infty}^{\infty} u(x, \Delta t)\psi(x + k\Delta x) \, dx$$

$$\tag{11.12}$$

$$= \frac{1}{\sqrt{4\pi\Delta t}} \int_{-\infty}^{\infty} \int_{-\infty}^{\infty} \psi(\eta)\psi(x + k\Delta x)e^{-\frac{(\eta-x)^2}{4\Delta T}} \, d\eta \, dx .$$

Denote by $a(f, g)$ the $(2J+1)$-vector $(a_{-J}, \ldots, a_J)$ determined by solving (11.10)–(11.12) with $\psi(x) = f(x), u(x) = g(x, \Delta t)$. Define operators $\mathcal{A} = \mathcal{A}(f, g)$ and $\mathcal{K}_{\Delta t}$ by

$$\mathcal{A}w = \sum_{j=-J}^{J} a_j \, w(x + j\Delta x),$$

$$\mathcal{K}_{\Delta t}w = \frac{1}{\sqrt{4\pi\Delta t}} \int_{-\infty}^{\infty} w(\eta) \, e^{-\frac{(\eta-x)^2}{4\Delta t}} \, d\eta .$$

The synergized method produces sequences ψ_n, u_{n+1} where $\psi_0 = \psi(x)$, $u_1 = u(x, \Delta t)$ and

$$\psi_{n+1} = \mathcal{A}(\psi_n, u_{n+1})\psi_n ,$$

$$\tag{11.13}$$

$$u_{n+2} = \mathcal{A}(\psi_n, u_{n+1})u_{n+1} .$$

One can verify that

$$u_{n+2} = \mathcal{K}_{\Delta t} \, \psi_{n+1} \, .$$

The parallelism in the algorithm consists of computing a pair of functions (ψ_n, u_{n+1}) (on a mesh) in parallel; the synergy is introduced through the determination of operators $\mathcal{A}(\psi_n, u_{n+1})$ which up-date the accuracy of the coefficients a_j.

The above scheme was introduced in [2]. In [1] computations are carried out in case $J = 1$ comparing the synergistic algorithm with the conventional (serial) algorithm where $a_0 = 0, a_1 = a_{-1} = \dfrac{1}{2}$, taking the initial data

$$u(x, 0) = e^{-x^2} \cos kx \qquad (k = 1, 2, \ldots, 8) \, .$$

It is found that s increases from a factor of 4 for $k = 1$ to a factor of 250 for $k = 8$. It would be interesting to find at which value of k breakeven $(s \geq s^*)$ occurs.

11.4   Open Questions

(i) Consider the problem (11.9), but use implicit schemes, i.e.,

$$U(x, t + \Delta t) - U(x, t) = \sum a_j U(x + j\Delta x, t + \Delta t)$$

Device a synergistic parallel algorithm with up-dated coefficients a_j.

(ii) Do the same for the wave equation.

11.5   References

[1] M.E. Henderson and W.L. Miranker, *Synergy in parallel algorithm*, June, 1987 (preprint).

[2] P.D. Gerber and W.L. Miranker, *Nonlinear Difference Schemes for linear partial differential equations*, Computing, 11 (1979), 197–211.

12

A Conservation Law Model for Ion Etching for Semiconductor Fabrication

In sections 12.1, 12.2 we shall describe a 2–dimensional mathematical model for ion etching which was presented on February 12, 1988 by David S. Ross from Eastman Kodak Company, and is based on his papers [1] [2]. In Section 12.3 several mathematical problems will be posed.

12.1   Etching of a Material Surface

A material with constant molecular density N occupies space $\{0 < y < y(x,t)\}$. It is etched at its surface $y = y(x,t)$ by bombardment with beam of ions with constant flux density ϕ (see Figure (12.1). The surface evolves according to [3]

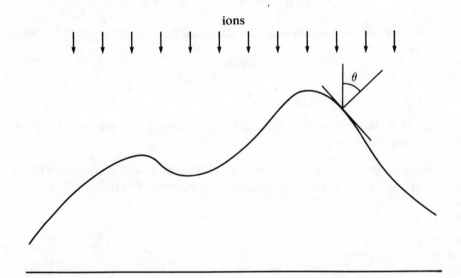

ions

θ

FIGURE 12.1.

$$y = -\frac{\phi}{N} \, S(\theta)$$

where $\theta = \arctan y_x(x,t)$ and $S(\theta)$ is a function which depends on the material. The function $f(p) = \frac{\phi}{N} S(p)$ is called the *sputtering function*; it is positive and uniformly bounded for $-\infty < p < \infty$; it has several inflection points (see Figure 12.2).

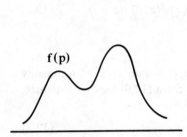

f(p)

Amorphous   material

(Silicon)

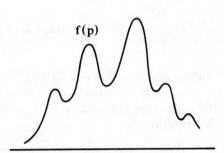

f(p)

Crystalline   material

(Gallium   Arsenide)

FIGURE 12.2.

The problem is to find the evolution of the etched surface, i.e., to solve the equation

$$y_t + f(y_x) = 0 \tag{12.1}$$

subject to

$$y(x,0) = y_0(x) . \tag{12.2}$$

If we differentiate (12.1) in x and set $y_x = p$, then the problem reduces to solving the conservation law

$$p_t + (f(p))_x = 0 \tag{12.3}$$

subject to

$$p(x,0) = p_0(x), \tag{12.4}$$

where $p_0(x) = y_0'(x)$.

Introducing the characteristic curves, i.e., the solutions of

$$\frac{dx}{dt} = f'(p(x,t)) \tag{12.5}$$

(assuming p is known) we compute, formally,

$$\frac{d\,p(x(t),t)}{dt} = p_t + p_x \frac{dx}{dt} = 0 ,$$

so that p is constant along any characteristic curve and thus, in (12.5), $f(p)$ is constant, i.e., the characteristic curves are straight lines with slopes determined by p_0. In general, these lines may meet and then a discontinuity of $p(x,t)$ is inevitable. Thus, even when $f(p)$ and p_0 are smooth functions, the solution $p(x,t)$ will develop discontinuities, in general. We must therefore define the concept of a solution to (12.3) so that it applies to discontinuous functions. Thus we are led to the "weak" formulation:

$$\int_0^\infty \int_{-\infty}^\infty (p\psi_t + f(p)\psi_x)\ dx dt + \int_{-\infty}^\infty p_0(x)\psi(x,0)\ dx = 0 \qquad (12.6)$$

for any test function $\psi \in C_0^1(\mathbf{R}^2)$.

If $x = s(t)$ is a smooth curve such that $p(x,t)$ is smooth (i.e., uniformly C^1) to the left of it and to the right of it, then from (12.6) we easily deduce, by integration,

$$\frac{ds(t)}{dt} = \frac{[f(p)]_s}{[p]_s} \qquad (12.7)$$

where

$$\begin{aligned} [p]_s &= p(s(t)+0,t) - p(s(t)-0,t) \\ [f(p)]_s &= f(p(s(t)+0,t)) - f(p(s(t)-0,t)); \end{aligned}$$

the relation (12.7) is called the *Rankine–Hugoniot condition*.

The definition (12.6) is not satisfactory, for there may exist many (nonphysical) solutions in the sense of (12.6); see, for instance, [4]. Thus one would like to incorporate a physically relevant criterion, called the *entropy condition*, into the concept of a solution. Since the introduction of discontinuities was motivated by the intersection of characteristics, a reasonable physical criterion might be: characteristics on either side of a discontinuity must meet as t increases. If $f(p)$ is convex this condition is sufficient to exclude all the physically unrealistic discontinuities; this condition (for f convex) is also equivalent to the Oleinik entropy condition [5]

$$\frac{p(x+a,t) - p(x,t)}{a} \leq \frac{E}{t} \qquad \forall\ a > 0,\ t > 0, x \in \mathbf{R} \qquad (12.8)$$

where E is a constant independent of x,t,a. It is well known [5] that for any $p_0 \in L^\infty$ there exists a unique entropy solution, i.e., a solution of (12.6), (12.8); further, $x \to p(x,t)$ is a function of bounded variation, for any $t > 0$; it is also piecewise smooth (e.g. DiPerna [6]).

In our case $f(p)$ is not convex, so we need the more stringent entropy condition, also due to Oleinik [7]:

Let $x = s(t)$ be any curve of discontinuity and denote by p^+ and p^- the limits of $p(x,t)$ as $x \to s(t)$ from the right–hand and left–hand sides, respectively. Then

$$\frac{f(p) - f(p^+)}{p - p^+} \leq \frac{ds}{dt} \leq \frac{f(p) - f(p^-)}{p - p^-}$$

$$(12.9)$$

for all p between p^+ and p^- .

One can incorporate (12.6) and (12.9) in the following:

Definition 12.1 *A bounded measurable function $p(x, t)$ on $\mathbb{R} \times \mathbb{R}^+$ is called an entropy solution of (12.3), (12.4) if for any convex functions $g(p)$ and a test function $\psi \in C_0^1(\mathbb{R}^2)$*

$$\int\int_{t>0} [g(p)\psi_t + F(p)\psi_x] \, dx dt + \int_{-\infty}^{\infty} g(p_0(x))\psi(x, 0) \, dx \geq 0$$

where $F(p) = \int^p f'(p) \, dg(p)$.

If p is piecewise smooth on both sides of a smooth curve $x = s(t)$, then (12.7) and (12.9) follow from Definition 12.1.

For existence and uniqueness of the entropy solution for non-convex f, see [7] [8] [9] [10] [11]; the flux function f need only be C^1. The solution may have a dense set of discontinuities [12].

The existence and uniqueness results for entropy solutions have been extended to a conservation law in several space variables by Volpert [13] (in BV class for the initial data and for the solution) and by Kruskov [14] (in L^∞ class).

There are various methods to prove existence in one space dimension. The most suggestive one is the viscosity method; solve the parabolic equation

$$p_t + (f(p))_x - \epsilon \, p_{xx} = 0 \qquad (12.10)$$

subject to (12.4), and show that a subsequence of solutions p_ϵ is convergent to an entropy solution. This approach and finite-difference variants of it have been used by several authors (e.g. [5] [7]); for a review and reference on other methods see [4; 304] [11].

12.2   Etching in Semiconductor Device Fabrication

The physical setting arises when one etches a material used in semiconductor device as shown in Figure 12.3.

Thus, in general, we are given two materials, the upper one with surface $y = y^u(x, t)$ and the lower one with surface $y = y^\ell(x, t)$, separated by a given interface $y = g(x)$, as shown in Figure 12.4; $g(x)$ is continuous and

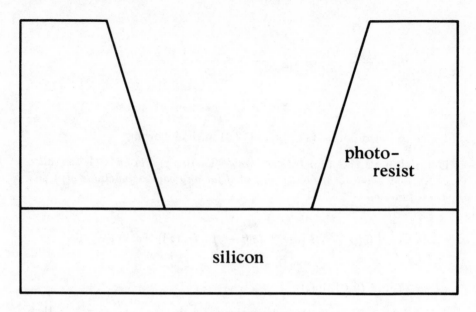

FIGURE 12.3.

piecewise smooth. The point $x_b(t)$ at the common boundary of the two surfaces is called the *material boundary*. The materials are characterized by different sputtering (or flux) functions, $f^u(p)$ for the upper material and $f^\ell(p)$ for the lower material. The upper material erodes in time and the rate of erosion should determine the speed of the material boundary $x_b(t)$. In the upper material,

$$y_t^u + f^u(y_x^u(x,t)) = 0 \tag{12.11}$$

and in the lower material,

$$y_t^\ell + f^\ell(y_x^\ell(x,t)) = 0 . \tag{12.12}$$

Since the joint surface $y = y(x,t)$ should be continuous at the material boundary, we have

$$y^u(x_b(t),t) = y^\ell(x_b(t),t) = g(x_b(t)). \tag{12.13}$$

Set $p^u(x,t) = y_x^u(x,t), p^\ell(x,t) = y_x^\ell(x,t)$. Then, from (12.11), (12.12),

$$
\begin{aligned}
p_t^u + (f^u(p^u))_x &= 0 , &\tag{12.14}\\
p_t^\ell + (f^u(p^\ell))_x &= 0 . &\tag{12.15}
\end{aligned}
$$

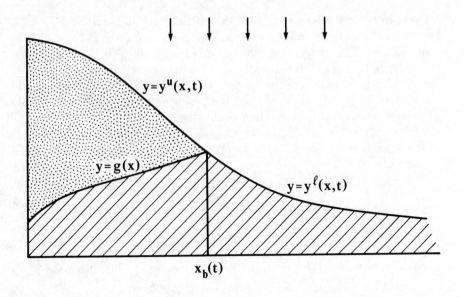

FIGURE 12.4.

Introduce the quantities

$$p_{\lim}^u(t) = \lim_{x \to x_b(t)} p^u(x, t),$$

$$p_b^u(t) = p^u(x_b(t), t)$$

and similarly $p_{\lim}^\ell(t)$, $p_b^\ell(t)$; $x_b(t)$ may not be differentiable at some points, and $p_{\lim}^u(t)$, $p_{\lim}^\ell(t)$ may not coincide with $p_b^u(t)$ and $p_b^\ell(t)$ respectively.

Equations (12.11), (12.12) hold in each material, up to the boundary, so that

$$y_t^u + f^u(p_b^u(t)) = 0 , \tag{12.16}$$

$$y_t^\ell + f^\ell(p_b^\ell(t)) = 0 \tag{12.17}$$

where $y^u = y^u(x_b(t), t)$ $y^\ell = y^\ell(x_b(t), t)$). Next, if we differentiate (12.13) we get

$$y_t^u + \frac{dx_b(t)}{dt} p_b^u(t)) = g'(x_b(t)) \frac{dx_b(t)}{dt} ,$$

$$y_t^\ell + \frac{dx_b(t)}{dt} p_b^\ell(t) = g'(x_b(t)) \frac{dx_b(t)}{dt} .$$

Comparing these equations with (12.14) and (12.15), we obtain

$$\frac{f^u(p_b^u(t))}{p_b^u(t) - g'(x_b(t))} = \frac{dx_b(t)}{dt} = \frac{f^\ell(p_b^\ell(t))}{p_b^\ell(t) - g'(x_b(t))} . \tag{12.18}$$

These equations describe the evolution of the material boundary.

The model of ion etching can thus be represented by (12.14), (12.15) and (12.18). This model does not incorporate effects such as ion reflection or redeposition of sputtered material. Therefore the evolution of the upper material is independent of the lower material. This suggests that one should first solve for p^u, then find $x_b(t)$ by the first equality in (12.18), and finally solve for p^ℓ using the information about $x_b(t)$.

In terms of a numerical scheme, suppose at time t we know $x_b(t)$, $p^u(x, t)$ for $x < x_b(t)$ and $p^\ell(x, t)$ for $x > x_b(t)$, and we wish to determine the correct values of

$$\frac{dx_b(t)}{dt}, \quad p_b^u(t), \quad p_b^\ell(t) \tag{12.19}$$

to be used in the next time–step. Then the schemes should be:

(i) Use $p_{\lim}^u(t)$ in order to determine $p_b^u(t)$.

(ii) Compute $dx^b(t)/dt$ from the first relation in (12.18), using the $p_b^u(t)$ obtained in (i).

(iii) Determine $p_b^\ell(t)$ from the second relation in (12.18), using the $dx_b(t)/dt$ computed in step (ii).

If this scheme can be shown to yield unique values of the quantities in (12.19), then numerical finite-difference schemes (such as in Ross [1]) can be set up in order to construct a numerical solution to (12.14), (12.15), (12.18) for given initial data. This however still leaves open questions of existence and uniqueness.

The proofs that (iii) yields a unique $p_b^\ell(t)$ and (i) yields a unique $p_b^u(t)$ are given in Ross [2], assuming that $f''(p)$ has a finite number of zeroes; it is based on physical considerations: For (iii) one uses the facts that $dx_b/dt \leq 0$ and that information about the material boundary must be propagated into the lower material. This implies that if $p_{\lim}^\ell(t)$ does not satisfy the equation

$$f^\ell(p) = \frac{dx_b(t)}{dt} \left(p - g'(x_b(t))\right) \tag{12.20}$$

then $p_b^\ell(t)$ is the unique solution of (20) such that

$$f(p_b^\ell(t)) - f(p) < \tfrac{dx_b(t)}{dt} \left(p_b^\ell(t) - p\right)$$

$$\text{for all} \quad p \quad \text{between} \quad p_b^\ell(t) \quad \text{and} \quad p_{\lim}^\ell(t); \tag{12.21}$$

if $p_{\lim}^\ell(t)$ satisfies (12.20) then we take $p_b^\ell(t) = p_{\lim}^\ell(t)$.

The quantity $p_b^u(t)$ is defined to be the largest number $\leq p_{\lim}^u(t)$ such that it satisfies

$$f^u(p) = \frac{dx_b(t)}{dt}(p - g'(x_b(t))) \tag{12.22}$$

and

$$\frac{f(p_b^u(t))}{p_b^u(t) - g'(x_b(t))} \; (p - g'(x_b(t)) \geq f(p) \qquad \forall \, p < p_{\lim}^u(t). \qquad (12.23)$$

This definition is partially based on the fact that information must be propagated from the material boundary into the upper boundary. But there is also an additional requirement involved in the derivation of (12.23), namely, characteristics should not initiate at the material boundary which diverge from the material boundary (for this would leave the solution undetermined in the region traced by these characteristics).

12.3 Open Problems

Consider a finite difference scheme based on the updates provided for in steps (i)–(iii) of section 12.2.

Problems. (1) Prove that the scheme is convergent to an entropy solution which satisfies (12.18) (in some weak sense).

(2) Is $x_b(t)$ piecewise smooth?

(3) Is the solution unique?

(4) How does $x_b(t)$ behave as $t \to \infty$?

Another possible approach to solving the problem (12.14), (12.15), (12.18) is by the viscosity method; it might be more convenient however to work with the Hamilton–Jacobi approach as in [15] [16]. Thus, if we set

$$w_1 = y^u(x,t), \quad w_2 = y^\ell(x,t), f_1 = f^u, \; t_2 = f^\ell, \quad s(t) = x_b(t)$$

then, for any $\epsilon > 0$, we wish to solve with

$$w_{1,t} - \epsilon w_{1,xx} + f_1(w_{1,x}) = 0 \quad \text{if} \quad x < s(t), \qquad (12.24)$$

$$w_1(x,t) = g(x) \qquad \text{on} \quad x = s(t), \qquad (12.25)$$

$$w_1(x,0) = w_{1,0}(x) \qquad \text{if} \quad x < s(0), \qquad (12.26)$$

and

$$w_{2,t} - \epsilon \, w_{2,xx} + f_2(w_{2,x}) = 0 \qquad \text{if} \quad x > s(t), \qquad (12.27)$$

with

$$w_2(x,t) = g(x) \qquad \text{on} \quad x = s(t), \qquad (12.28)$$

$$w_2(x,0) = w_{2,0}(x) \qquad \text{if} \quad x > s(0). \qquad (12.29)$$

However an additional condition on the free boundary is needed in order to solve the problems (12.24)–(12.29). David Ross suggested that an appropriate condition might be

$$\frac{\partial w_1}{\partial x} = -\phi(\epsilon) \quad \text{on} \quad x = s(t) - 0 \tag{12.30}$$

where $\phi(\epsilon) \to \infty$ as $\epsilon \to \infty$; this is based on the fact that in the derivation of (12.23) (for $g(x) = -cx$, $w_1(x,0) = bx$ with $c > 0, b > 0$) the function $1/p$ changes sign across the free boundary and thus, formally, $p^u = -\infty$ on the free boundary. The problem (12.24, (12.25), (12.26), (12.29) of the type of a Stefan problem for supercooled water.

Problem (5). Can the above approach be carried out, and, if so, does the solution satisfy the "viscosity solution" characterization of Crandall, Evans and Lions [17]?

In etching problems one can change the sputtering function f by changing the chemical nature of the ions. Thus f becomes a function $f(p,c)$ where c is a parameter; for instance, $f(p,c) = f_0(p-c)$ for some function f_0. Typical initial shape and desired final shape are diagrammed in Figure 12.5

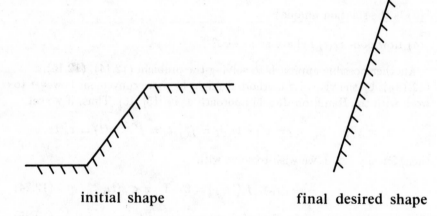

initial shape   final desired shape

FIGURE 12.5.

Problem (6). Choose $c(t)$ such that the final shape is as close as possible (say in L^2 sense) to the desired final shape.

Notice that in this model the ions are still taken to be uniformly distributed at each time, but their constitution changes in time. The above model can be extended to $f = f(p, c_1, c_2, \ldots, c_n)$.

12.4 REFERENCES

[1] D.S. Ross, *Ion etching: An application of the mathematical theory of hyperbolic conservation laws*, J. Electrochemical Society, 10 South Main Street, Pennigton, N.J, to appear .

[2] D.S. Ross, *Evolution of material boundaries under ion bombardment*, J. Electrochemical Society, 10 South Main Street, Pennington N.J, to appear.

[3] R. Smith and J.M. Walls, Philosophical Magazine A, vol. 42 (1980), p. 235.

[4] J. Smoller, *Schock Waves and Reaction Diffusion Equations*, Springer-Verlag, New York, 1980.

[5] O. Oleinik *Discontinuous solutions of nonlinear differential equations*, Amer. Math. Soc. Translations, Sec. 2, 26 (1957), 95–172.

[6] R. DiPerna, *Singularities of solutions of nonlinear hyperbolic systems of conservation laws*, Archive Rat. Mech. Anal., 60 (1975), 75–100.

[7] O. Oleinik, *Uniqueness and a stability of the generalized solution of the Cauchy problem for a quasilinear equation*, Amer. Math. Soc. Transl., Sec. 2, 33 (1964), 285–290.

[8] D. Ballou, *Solutions to nonlinear hyperbolic Cauchy problems without convexity conditions*, Trans. Amer. Math. Soc., 152 (1970), 441–460.

[9] C.C. Wu, *On the existence and uniqueness of generalized solutions of the Cauchy problem for quasilinear equations of the first order without convexity conditions*, Acta Math. Sinica 13 (1963), 515–530 (Chinese Math.–Acta, 41 (1964), 561–577).

[10] C.M. Dafermos, *Polygonal approximations of solutions of the initial value problems for a conservative law*, J. Math. Anal. Appl., 38 (1972), 33–41.

[11] C.M. Dafermos, *Characteristics in hyperbolic conservation laws, a study of the structure and asymptotic behavior of solutions in* "Nonlinear Analysis and Mechanics" Heriot–Watt Symposium I, Ed. R.J. Knops, Research Notes in Mathematics II 17, Pitman, London, 1977, 1–58.

[12] D. Ballou, *Weak solutions with 9 dense set of discontinuities*, J. Diff. Eqs., 10 (1971), 270–280.

[13] A. Volpert, *The spaces BV and quasilinear equations*, Mat. Sbornik, 73 (1967), 255–302 (English transl. in Math. USSR Sb., 2 (1967), 225–267).

[14] S. Kruskov, *First order quasilinear equations with several space variables*, Mat. Sbornik, 123 (1970), 228–255 (English transl. in Math. USSR Sb., 10 (1970), 217–273).

[15] W.H. Fleming, *The Cauchy problem for degenerate parabolic equations*, J. Math. Mech., 13 (1964), 987–1008.

[16] A. Friedman, *The Cauchy problem for first order partial differential equations*, Indiana Univ. Math. J., 23 (1973), 27–40.

[17] M.G. Crandall, L. C. Evans and P.L. Lions, *Some properties of viscosity solutions of Hamilton–Jacobi equations*, Trans. Amer. Math. Soc., 282 (1984), 487–502.

13

Phase Change Problems with Void

13.1   The Problem

A problem on a phase change related to thermal energy storage in Manned Space Station is being studied by D.G. Wilson, J.B. Drake and R.E. Flanery at Oak Ridge National Laboratory:

"It is intended that the manned space station satisfy a considerable portion of its power requirements with solar energy. The station will orbit the earth in about ninety minutes and spend about two thirds of each orbit in sunlight and about one third in the earth's shadow. Under these conditions it will be necessary to store thermal energy during the station's exposure to the sun and to retrieve it during the transit through the earth's shadow. Several systems have been proposed for accomplishing this. We describe in rough outline one such system and a problem associated with modeling its performance.

The system considered consists of a solar collector lined with small metal canisters filled with a high temperature phase change material (PCM), lithium fluoride salt (Figure 13.1).

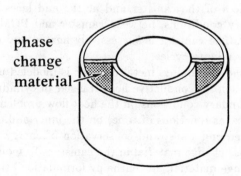

containment canister

phase change material

FIGURE 13.1.

The canisters are small enough to fit comfortably in the palm of one's hand

and there are a hundred or more of them. A heat transfer fluid, an inert gas such as helium or neon, circulates through pipes that pass through the metal canisters and carries heat away to turbines, generators, etc. The continual remelting and refreezing of the PCM smears out the delivery of the solar energy to the transfer fluid and hence to the heat engines beyond. The motivation for using a PCM based thermal energy storage system is that a properly sized such system can store and deliver energy over a narrow temperature range near the melting point of the PCM thus avoiding extreme temperature variations.

A multitude of problems must be solved to design a satisfactory system. The size of the solar collector must be matched with the power requirements of the station. The capacity of the energy storage system must be such that the PCM just about completely melts during the insolation period and just about completely freezes during the dark period. (Otherwise the advantage of the PCM is lost.) The canister material and design must be adequate to withstand the frequent meltings and freezings of the PCM, and the resulting mechanical stresses, for many cycles.

The problem that we have addressed, whose solution will contribute to a stress analysis of the canisters and a prediction of the performance of the storage system, is the modeling of the thermal and fluid flows in a representative canister. There will be fluid flow in the liquid PCM, even in a microgravity environment, because there is a significant difference in density between solid and liquid lithium fluoride. This causes the formation of one or more voids on freezing and, of course, the disappearance of voids on remelting.

The problem consists of a system of coupled, mildly nonlinear, partial differential equations for heat flow and fluid flow. The geometry of the problem is as follows. A right circular cylindrical canister is filled with a phase change material. An enclosed cylindrical pipe runs down the center of the canister. Boundary conditions are supplied at the inner and outer cylindrical surfaces of the canister and at the end faces of the canister. Internal boundary conditions, between canister and PCM, insure conservation of energy. The canister is successively heated and cooled from the outside in ninety minute cycles.

Heat flow in the containing, metal canister is modeled using the partial differential equation for conductive heat transfer in cylindrical polar coordinates. The boundary conditions on the heat flow problem in the canister are: imposed flux (as functions of time) on the inner and outer cylindrical surfaces, insulated end walls, and conservation of energy at the interface between the phase change material in the canister. To model the heat flow in the phase change material, an "enthalpy formulation" that permits easy treatment of the successive melting and freezing cycles is used. A convection term is included to account for the heat transfer caused by bulk movement of the liquid phase change material. The boundary conditions on the heat flow problem in the phase change material are just conservation of energy in communicating with the walls of the canister. The movement of the liquid is modeled using a weak formulation of the incompressible Navier Stokes equations that is consistent with the enthalpy formulation of the heat transfer equations. The density of the solid phase change material is about 1.25 times that of liquid material, and thus a vapor filled void forms when material freezes. The phase transition region is treated as a porous medium that inhibits fluid flow but also introduces the density change that acts as a source for the fluid flow. The surface of the void is a moving boundary (or free surface) whose determination is part of the problem. Effects of surface curvature, surface tension and Marangoni forces at this free surface are included in the fluid flow equations. The boundary conditions on the fluid flow problem in the phase change material are: "no slip," i.e., all velocity components are zero, at the walls of the canister, and conservation of mass and momentum at the void surface."

13.2   The Void Problem in 1–Dimension

Solidification and melting of a material are commonly modeled as a Stefan problem. Thus if a liquid region $D(t)$ (which changes with time t) is surrounded by solid at the melting temperature T_0, and if the boundary of $D(t)$ is given by an equation $\Phi(x,t) = 0$, then the temperature T in the fluid satisfies the heat equation

$$T_t - \Delta T = 0 \quad \text{in} \quad D(t) , \tag{13.1}$$

and the interface conditions

$$T = 0, \quad k\nabla_x\Phi \cdot \nabla_x T = \Phi_t \quad \text{on} \quad \Phi(x,t) = 0 , \tag{13.2}$$

where k is a positive constant.

When the solid's temperature is smaller than T_0, then equation (13.1) is satisfied also in the region $E(t)$ occupied by the solid, and the interface conditions are

$$T^{\pm} = 0, \quad k_1 \nabla_x\Phi \cdot \nabla_x T^+ - k_2\nabla_x\Phi \cdot \nabla_x T^- = \Phi_t \quad \text{on} \quad \Phi = 0 , \tag{13.3}$$

where T^+ (T^-) is the restriction of T to the liquid (solid).

In order to set up the Stefan problem in full details, one must add boundary conditions on T at the fixed portions of the liquid and solid boundaries and also prescribe initial values $T(x,0)$. The Stefan problems corresponding to (13.2) and (13.3) are called, respectively, the one–phase and two–phase Stefan problems.

Existence and uniqueness of a weak solution for the Stefan problems are well known [1]; further, T is continuous [2] [3]. For the one–phase Stefan problem the free boundary (i.e., the interface) is known to be a smooth

C^∞ surface, for some restricted classes of initial and boundary data [4] (see also [1]); for arbitrary data, the free boundary is a set of measure 0 and the second condition in (13.2) holds only in some weak sense. For the two–phase Stefan problem, much less is known about the free boundary (defined to be the set $\{T = T_0\}$).

What is ignored in the above model is the fact that the solid and liquid have different densities, a fact which causes vapor–filled bubbles or voids to form in the liquid. This problem was examined by Solomon et al [5]. They study the thermal performance of a latent–heat thermal energy storage system for a space–station power cycle. The phase change material (PCM) they studied is lithium fluoride, whose density (as mentioned in section 13.1) increases upon freezing by about 25%. Since the PCM container must provide storage for the material in its liquid phase, solidification must be accompanied by formation of one or more voids. A treatment of the initial formation of voids seems very difficult; cf. Alexiades et al [6]. But even the evolution of macroscopic voids is not easy. Yet such information is important for the proper design of space power system and its PCM storage module.

Wilson and Solomon [7] have studied a very simple one–dimensional two–phase Stefan problem where the solid lies between the liquid and the void; see Figure 13.2. The process takes place in $\{x > 0, \, t > 0\}$, and along the free boundary $x = s(t)$ the usual Stefan conditions hold.

In the solid T satisfies $T_t + \gamma s' \, T_x = T_{xx}$ (rather than the usual heat equation). Because the density ρ_L of liquid is smaller than the density ρ_S of solid, the conservation of mass dictates that the void must occupy the interval $0 < x < \gamma s(t)$ where

$$\gamma = 1 - \rho_L/\rho_S .$$

Assuming Newton's cooling law between the solid and void interface, they are led to the boundary condition

$$-T_x(\gamma s(t) + 0, t) = a(\widetilde{T} - T(\gamma s(t) + 0, t)) \qquad (a > 0).$$

If $s(0) = 0$ and $T(x, 0) = $ const., then an explicit similarity solution can be obtained, with $s(t) = $ const. $\sqrt{t}$; see [7].

Notice that the precise location of the void's boundary, in the above model, is completely determined by the conservation of mass. In space dimension ≥ 2 this is no longer the case, and the velocity field of the void's boundary must be determined by dynamic conditions (in addition to the conservation of mass).

George Wilson presented this void problem in a talk on February 19, 1988. He posed the question: how to determine the evolution of the void? In subsequent discussions with him we have outlined a procedure which determines the evolution of *one* void, contained in the liquid. This is explained in the next section.

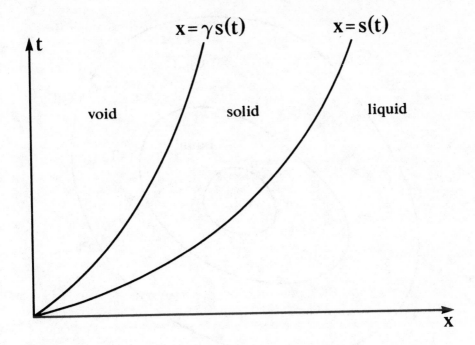

FIGURE 13.2.

13.3 A Scheme to Solve the Void Problem

Although the scheme can be implemented in principle, in any number of dimensions, its execution will be simpler in the case of two space dimensions. Since the apparatus used for space–station is a cylinder, the two–dimensional problem is of physical relevance. We wish to develop a two–dimensional model as shown in Figure 13.3.

On the outer boundary S_0 the temperature T is given, $T = k$ and $k < T_0$ (T_0 is the solidification temperature).

We shall consider the problem as composed of two free boundary problems: the Stefan problem in the region between S and S_0 (with free boundary Γ), and the bubble problem in ideal incompressible fluid in the region enclosed by Γ, with free boundary S. The two problems are coupled.

We shall describe an iterative scheme to be carried out in time intervals $t_0 \leq t \leq t_1, \quad t_1 \leq t \leq t_2, \ldots, \quad t_{n-1} \leq t \leq t_n, \ldots$, where $t_j - t_{j-1} = \delta$ and δ is small.

Step I_j Solve the two–phase Stefan problem in the cylindrical domain bounded by the interior lateral boundary $S \times (t_j, t_{j+1})$ and the exterior lateral boundary $S_0 \times (t_j, t_{j+1})$, with initial free boundary Γ. Here we take

$$T = k \quad \text{on} \quad S_0 \times (t_j, t_{j+1}) , \tag{13.4}$$

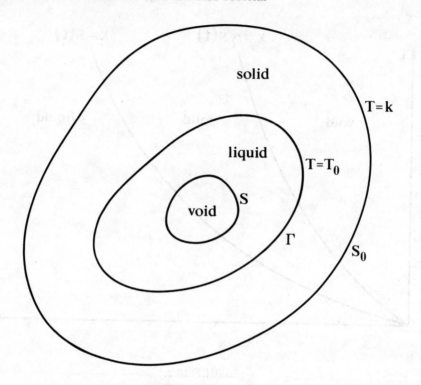

FIGURE 13.3.

$$\frac{\partial T}{\partial \nu} + h(T - \hat{T}) = 0 \qquad \text{on} \qquad S \times (t_j, \, t_{j+1}) \qquad (13.5)$$

where $\hat{T}$ is the temperature in the void's boundary. Since we do not know what this temperature is, it is reasonable to take

$$\hat{T} = \frac{1}{|S|} \int_S T \, .$$

Thus the boundary condition (13.5) is slightly more complicated than usual.

Step II_j We shall assume that the fluid is ideal, incompressible and irrotational (Irrotationality is justified in space–station, since gravity is negligible.). Then the velocity potential u satisfies

$$\Delta u = 0 \quad \text{for} \quad x \in \Omega_t \qquad (t_j \le t \le t_{j+1}) \qquad (13.6)$$

where Ω_t is the region bounded by S and Γ_t, the evolving free boundary obtained in Step I_j. Since the fluid does not leave Ω_t through Γ_t, we also have

$$\frac{\partial u}{\partial \nu} = 0 \qquad \text{on} \qquad \Gamma_t \, . \qquad (13.7)$$

As for the boundary condition on the void's boundary, if we denote this boundary by $r = R(\theta, t)$ (using polar coordinates (r, θ) centered about a point in the void) then the equation of continuity yields

$$\frac{\partial u}{\partial r} - \frac{1}{r^2} \frac{\partial u}{\partial \theta} \frac{\partial R}{\partial \theta} = \frac{\partial R}{\partial t} \quad \text{on} \quad r = R(t), \tag{13.8}$$

and Bernoulli's law gives

$$|\nabla u|^2 + \frac{c\,\hat{T}}{|V|^\delta} = c_0 \quad \text{on} \quad r = R(t) \tag{13.9}$$

where $|V|$ is the volume of the void and c_0 is a given constant; note that $c\hat{T}/|V|^\gamma$ is the pressure p of the vapor–filled bubble or void.

As we solve step II_j we obtain $u(x, t)$ and $S_t \equiv \{r = R(\theta, t)\}$ for $t_j \leq t \leq t_{j+1}$. Next, with $S_{t_{j+1}}, u(x, t_{j+1})$ at hand, we can continue with step I_{j+1}, then II_{j+1}, etc.

The Stefan problem solved in step I_j is a two–phase problem, and thus almost nothing is known about its free boundary, except in small times. For small times the free boundary is smooth if the data are smooth (see [8] [9]). If we restrict ourselves to one–phase problem with $T \equiv T_0$ in the liquid, then more is known on the free boundary problem. As for the 2–dimensional free boundary bubble problem to be solved in step II_j, a version of this problem was recently studied by Xinfu Chen and Avner Friedman, namely, the fluid occupies the entire space outside the bubble, and gravity may be included. They proved that a smooth unique solution exists for a small time.

Thus it appears that the procedure outlined above may yield a smooth solution as $\max(t_{j+1} - t_j) \longrightarrow 0$, for a small time interval.

Problems. (1)   Prove that for small $\tilde{t}_0$ the above scheme can be carried out for all $t_n < \tilde{t}_0$, and that it converges (as $\max(t_{j+1} - t_j) \to 0$) to a classical solution of the void problem.

(2)   Use the above scheme to study numerically the evolution of the void, until it intersects the liquid–solid interface; here numerical schemes for both the two–phase Stefan problem and the bubble problem need to be used.

For small voids one should include surface tension effects in the equation (13.9).

13.4   REFERENCES

[1] A. Friedman, *Variational Principles and Free–Boundary Problems*, Wiley & Sons, New York, 1982.

[2] L.A. Caffarelli and A. Friedman, *Continuity of the temperature in the Stefan problem*, Indiana Univ. Math. J., 28 (1979), 53–70.

[3] L. Caffarelli and L.C. Evans, *Continuity of the temperature for the two phase Stefan problem*, Arch. Rat. Mech. Anal., 81 (1983), 199–220.

[4] A. Friedman and D. Kinderlehrer, *A one–phase Stefan problem*, Indiana Univ. Math. J., 24 (1975), 1005–1035.

[5] A.D. Solomon, M. Morris, J. Martin and M. Olszewski, *The development of a simulation code for a latent heat thermal energy storage system in a space station*, Report No. ORNL–6213, Oak Ridge National Laboratory, 1986.

[6] V. Alexiades, A.D. Solomon and D.G. Wilson, *The formation of a solid nucleus in supercooled fluid*, Report No. ORNL–6280, Oak Ridge National Laboratory, 1986.

[7] D.G. Wilson and A.D. Solomon, *A Stefan–type problem with void formation and its explicit solution*, IMA J. Appl. Math., 37 (1986), 67–76.

[8] A.M. Marimanov, *On a classical solution of the multidimensional Stefan problem for quasilinear parabolic equations*, Mat. Sbornik, 112 (1980), 170–192.

[9] Ei-Ichi Hanzawa, *Classical solution to the Stefan problem*, Tohoku Math. J., 33 (1981), 297–335.

14

Combinatorial Problems Arising in Network Optimization

14.1 General Concepts

A *graph* G consists of a pair (V, E) where $V = V(G)$ is a finite set of points, called *vertices* or *nodes*, and $E = E(G)$ is a set of unordered pairs of vertices, called *edges* or *links*. Two vertices are said to be *adjacent* if there is an edge joining them. The *degree* of a vertex is the number of adjacent vertices. If every vertex has the same degree d then the graph is called *regular of degree* d; examples of regular graphs are given in Figure 14.1.

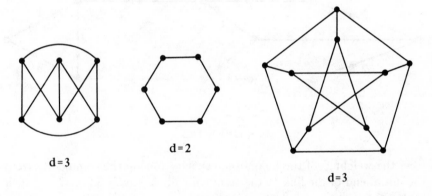

FIGURE 14.1.

If G is a graph with vertex set $\{v_1, v_2, \ldots, v_n\}$, we define the *incidence matrix* $A = (a_{ij})$ by a_{ij} = number of edges joining vertex v_i to vertex v_j. Thus $a_{ij} = 1$ $(= 0)$ if v_j is adjacent (not adjacent) to v_i, $a_{ij} = a_{ji}$, and $\sum_{j=1}^{n} a_{ij}$ is the degree of vertex v_i. If G is a regular graph of degree d then the principal eigenvalue of A is d (with $(1, \ldots, 1)$ as an eigenvector).

If each edge is given a direction then the graph is called a *directed graph*, or *digraph*. We then define the *indegree* (*outdegree*) of a vertex v as the sum of all directed edges ending (initiating) at v.

Let D be a digraph and ψ a nonnegative function (called *capacity* or *weight*) defined on the directed edges of D. The *inflow* (*outflow*) at a vertex v is defined as the sum $\Sigma\psi(\ell_i)$ taken over all directed edges ℓ_i ending (initiating) at v (If $\psi \equiv 1$ then inflow = indegree and outflow = outdegree).

A network $N = (D, \psi, s, t)$ consists of a digraph, a capacity function ψ, a vertex s, called *source*, with indegree 0 and a vertex t, called *terminal*, with outdegree 0; an example is given in Figure 14.2.

Graphs and networks are used to describe models in communication, transportation, distribution of commodities, scheduling etc. Telephone lines, traffic patterns and electrical networks are some of the most visible application areas. Electrical network (or a circuit) is a network where each directed edge ℓ has a prescribed resistance or impedance. A standard problem is to find the current in each link when a voltage potential is applied between two terminal points. Here the well known Kirchoff law asserts that at each node other than the source and terminal the inflow of current is equal to the outflow of current.

As another example consider a flow problem for commodities, illustrated in Figure 14.2.

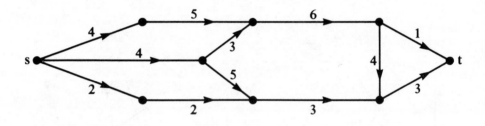

FIGURE 14.2.

Here the weight ψ of the link is the capacity to ship the commodity from the initial end of the link to the terminal end. A *flow* is a function ϕ such that $\phi \leq \psi$ and the ϕ-inflow is equal to the ϕ-outflow at each vertex other than s and t. The *value* of the flow is defined as the ϕ-outflow at s (which is the same as the ϕ-inflow at t).

A *cut* C is a subset of directed edges such that each path from s to t includes a directed edge from C. The *capacity* of C is the sum of the capacities of its directed edges If the capacity is minimum then C is called a *minimum cut*. A well known result of Ford and Fulkerson in the following:

Max-flow min-cut theorem: In any network the value of the maximum value of the flow is equal to the capacity of the minimum cut.

The literature on graphs and networks, being motivated by many important applications, is extensive and fast growing. We refer to a basic book by Wilson [1] and to a collection of review articles in [2].

On February 26, 1988 Fan R.K. Chung from Bell Communications Research presented some problems arising in network optimization. This talk is reviewed in the sequel.

14.2 Diameter Estimation

In a graph G, the distance between any two vertices u and v is defined by

$$d(u,v) = \text{ smallest number of edges in paths connecting } u \text{ to } v \text{ .}$$

The diameter $D(G)$ of G is defined by

$$D(G) = \max\{d(u,v) \; ; \quad u,v \quad \text{in} \quad V(G)\} \text{ .}$$

In communication network one is interested in constructing graphs such that they have

> many vertices,
> few edges,
> small diameter,
> small maximum degree; some of these goals are

in conflict with each other. After setting an objective function, one faces the mathematical question of evaluating it.

An example of an objectivity function is

$$n(k,D) = \max\{|V(G)|; \quad \text{diam } G \leq D \text{ , max . degree } G \leq k\} \text{ .} \tag{14.1}$$

Problem. Determine (or estimate) $n(k,D)$.

It is easy to check that

$$n(k,D) \quad \leq \quad 1 + k + k(k-1) + \ldots + k(k-1)^D$$

$$\tag{14.2}$$

$$\equiv \quad n_0(k,D) \qquad \text{(the Moore bound).}$$

Equality holds in just a few cases (such as when $D = 1$, G is a $(k+1)$-clique ; $k = 2$, G is a $2D+1$ cycle). It follows that

$$D \geq \log_{k-1} n - \frac{2}{k} \text{ .}$$

Bollobás and de la Vega [3] proved that a random k-regular graph has diameter

$$\log_{k-1} n + \log_{k-1} \log n + c$$

with probability close to 1, where c is a positive constant (at most 10). It follows that a regular random graph with degree k and diameter D has (with high probability) at least

$$(k-1)^{D-1}/(2k \log k - 1) \tag{14.3}$$

vertices, which is fairly close to the Moore bound. The probabilistic proof does not provide an actual construction of algorithm.

Problem (1). Find an explicit construction which yields the lower bound for $n(k, D)$ as in (14.3).

So far there are only known explicit algorithms of graphs with

$$n(k, D) = \left[\frac{k}{2}\right]^D \; ;$$

this number is off by a factor of 2^D from the Moore bound; for more details see [4].

14.3 Reducing the Diameter

Think of the nodes of a graph as computers or processors. By adding an edge E (or t edges) we can reduce the diameter of the graph, thereby improving the communication and reducing the amount of necessary information flow. How to do it optimally? A lower bound on the diameter of the graph $G \cup tE$ is given by

Theorem 14.1 *(F.R.K. Chung and M.R. Garey [5])*

$$D(G \cup tE) \geq \frac{D(G)}{t+1} - 1 , \tag{14.4}$$

and the minimum is achieved when G is a path P_n, i.e., is a graph with vertices $1, 2, \ldots, n$ and edges $(1, 2), (2, 3), \ldots, (n-1, n)$.

Setting

$$f(t, d) = \min_{\substack{G \\ D(G)=d}} D(G \cup tE) ,$$

(14.4) means that

$$f(t, d) \geq \frac{d-t}{t+1} .$$

Estimates on $D(P_n \cup tE)$ (i.e., on $f(t, d)$ with $n-1 = d$) have been obtained by Schoone, Bodlaender and van Leeuwen [6] and further improved by Kerjouan [7]:

$$\frac{d+t-3}{t+1} \leq f(t, d) \leq \frac{d+t}{t+1} \quad \text{for } t \quad \text{even}$$

$$\frac{d+2t-4}{t+1} \leq f(t, d) \leq \frac{d+t}{t+1} \quad \text{for } t \quad \text{odd.}$$

Problem (2). Determine the exact value of $f(t, d)$.

It is conjectured that

$$f(t, d) \;=\; \frac{d+t}{t+1} \qquad \text{if } t \text{ is even}$$

$$=\; \frac{d + 2t - 4}{t+1} \qquad \text{if } t \text{ is odd.}$$

Some examples are given in Figure 14.3.

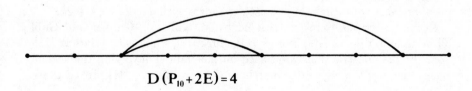

$$\mathbf{D\,(P_{10} + 2E) = 4}$$

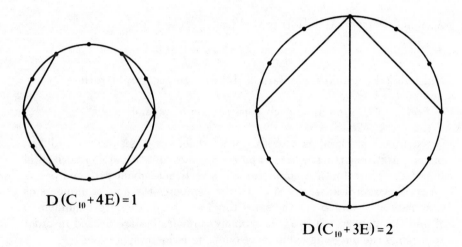

$$\mathbf{D\,(C_{10} + 4E) = 1}$$

$$\mathbf{D\,(C_{10} + 3E) = 2}$$

FIGURE 14.3.

14.4 Expander Graphs

The idea of an expander graph is that for any subset X of nodes the number $N(X)$ of its adjacent nodes should satisfy

$$N(X) \geq \alpha|X| \qquad (\alpha > 0); \tag{14.5}$$

suitable restrictions are to be placed on the size of $|X|$ and on α. Expanders are important in communication; they ensure that there will be no congestion. Thus many communication networks are designed as expander graphs.

We shall deal here with a specific but important case where G is a regular graph of degree d.

Let k be a positive parameter; we require that for any subset X with $|X| = [\frac{n}{k}]$

$$N(X) \geq n - [\frac{n}{k}] .\tag{14.6}$$

Such a graph will be called an *expander*. Expanders exist only if $d = d(k)$ is large enough, namely, $d \geq k \log k$. On the other hand by a probabilistic proof one can show that there exist expanders with $d \leq 2k \log k + 2k$.

Expander graphs were constructed by Margulis [8], Gabber and Galil [9], Alan and Milman [10], Tanner [11], Buck [12], Phillips and Sarnak [13]. Some of the constructions are based on inequalities involving the second eigenvalue λ of the adjacency matrix of G. Thus Tanner [11] proved that

$$N(X) \geq \frac{d^2 \, |X|}{(d^2 - \lambda^2)\frac{|X|}{n} + \lambda^2} \, , \qquad n = |V(G)|;$$

Alon and Boppana (see [14]) proved that

$$\liminf_{n \to \infty} \lambda \geq 2\sqrt{d-1} \, ,$$

and Lubotzky, Phillips and Sarnak [13] constructed graphs with $\lambda \leq 2\sqrt{d-1}$.

Problem (3). Find an explicit construction of expanders with $d \leq 2k \log k + 2k$.

Expanders are used as a tool to prove theorems. Consider the problem of comparative sorting, where numbers are moved in n parallel horizontal lines (see Figure 14.4). A comparator is a vertical segment with end points on two horizontal lines, say A and B. It compares the pair of numbers as they pass at A and B (at the same time) and sends the larger number to A and the smaller one to B. How many comparators are needed in order to arrange the outgoing columns of numbers in increasing order?

M. Ajtai, Y. Komlos and E. Szemerédi [15] [16] proved that $n \log n$ comparators are sufficient.

Expanders were used by P. Feldman, J. Friedman and N. Pipenger [17] in studying non blocking networks. Alon and Milman [10] proved that for a d-regular graph G,

$$D(G) \leq 2\sqrt{\frac{2d}{\lambda}} \, \log_2 n \quad ,$$

and Chung [18] proved that

$$D(G) \leq \frac{\log(n-1)}{\log(d/\lambda)} \quad ;$$

their proofs are based on expander graphs.

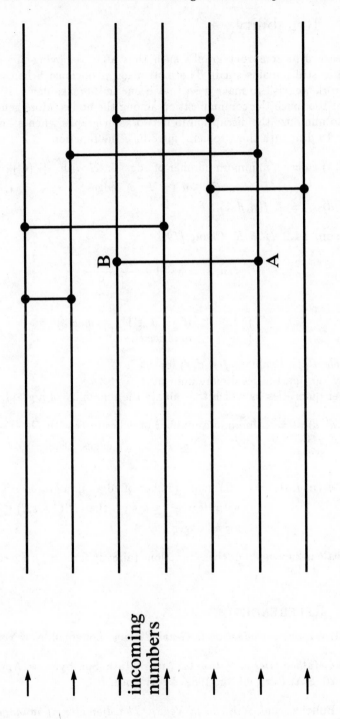

FIGURE 14.4.

14.5   Reliability

One wishes to construct graphs such that after removing any edge the diameter still remains small. The motivation is obvious: When we design a network we wish to make it so that if one link breaks down, it does not disrupt too much the communication among the nodes. More generally we wish to minimize the disruption in networks or graphs when s links break down. To deal with this problem, introduce the function

$$f(n, d, s) \quad = \quad \text{minimum number of edges in } G \text{ with} \quad |V(G)| = n \quad \text{such that}$$
$$D(G - sE) \leq d \quad \forall \quad s \quad \text{edges},$$
$$f(n, d) \quad = \quad f(n, d, 1) \ .$$

Theorem 14.2 *(F.R.K. Chung [19])*

$$f(n, d) = n - 1 + \frac{n - 1 - \epsilon}{\lceil \frac{2}{3} d \rceil}$$

where

$$\epsilon = \left\{ \begin{array}{ll} 1 & \quad if \quad\quad d \equiv 1 \quad (mod\ 3) \\ 0 & \quad otherwise \ . \end{array} \right. \tag{14.7}$$

Problem (4). Evaluate $f(n, d, s)$ for $s \geq 1$.
Only very partial results are known.
Other quantities by which to evaluate the reliability of a graph are

$$g(n, d, d', s) \quad = \quad \text{minimum number of edges in } G \text{ with } D(G) \leq d \text{ such that}$$
$$D(G - sE) \leq d' \quad \forall \quad s \quad \text{edges} \quad E,$$

$$g_k(n, d, s) \quad = \quad \text{minimum number of edges in a graph}$$
$$\text{with degree} \quad \leq k \text{ such that } D(G - sE) \leq d$$
$$\text{for all edges} \quad E \ .$$

Very little is known about these functions (already for $s = 1$); see references in [19].

14.6   REFERENCES

[1] R.J. Wilson, *Introduction to Graph Theory*, Longman, New York, 1985.

[2] *The Mathematics of Networks*, Amer. Math. Soc. Sympos. Appl. Math. vol. 26, S.A. Burr, editor, 1982, Providence, R.I.

[3] B. Bollobás and W.F. de la Vega, *The diameter of random graphs*, Combinatorica, 2 (1982), 125–134.

[4] F.R.K. Chung, *Diameters of graphs: old problems and new results*, preprint.

[5] F.R.K. Chung and M.R. Garey, *Diameter bounds for altered graphs*, J. of Graph Theory, 8 (1984), 511–534.

[6] A. Schoone, H. Bodlaender and J. van Leeuwen, *Diameter increase caused by edge deletion*, J. of Graph Theory, to appear.

[7] R. Kerjouan, *Arte-Vulnerabilite du diametre dans les reseaux d' interconnexion*.

[8] G. A. Margulis, *Explicit constructions of concentrators*, Problemy Inf. Trans., 9 (1973), 325–332.

[9] O. Gabber and Z. Galil, *Explicit constructions of linear-sized super-concentrators*, J. Compu. System Sci., 22 (1981), 407–420 .

[10] N. Alon, and V.D. Milman, λ_1 *isoperimetric inequalities for graphs and superconcentrators*, J. Combin. Theory, Ser. B, 38 (1985), 73–88.

[11] R.M. Tanner, *Explicit concentrators from generalized N-gons*, SIAM J. Alg. Disc. Methods, 5 (1984), 287–293.

[12] M.W. Buck, *Expanders and diffusers*, SIAM J. Alg. Disc. Methods, to appear.

[13] A. Lubotzky, R. Phillips, P. Sarnack, *Explicit Expanders and the Ramanujan Conjectures*, Proc. 18th ACM Symp. on Theory of Com., (1986), pp. 240–246.

[14] N. Alon, *Eigenvalues and expanders*, Combinatorica, 6 (1986), 83–96.

[15] M. Ajtai, J. Komlós, and E. Szemerédi, *An $0(n \log n)$ Sorting Network*, Proc. 15th ACM Symp. on Theory of Computing, (1983).

[16] M. Ajtai, J. Komlós and Szemerédi, *Sorting in $C \log N$ parallel steps*, Combinatorica 3, (1983) 1–19.

[17] P. Feldman, J. Friedman, N. Pipenger, *Non-Blocking Networks*, Proc. of 18th ACM Symp. on Theory of Computing, May (1986), pp. 247–254.

[18] F.R.K. Chung, *Diameters and eigenvalues*, to appear.

[19] F.R.K. Chung, *Graphs with small diameter after edge deletion*, to appear

15

Dynamic Inversion and Control of Nonlinear Systems

In Dynamic Inversion one wishes to use the available control in order to change the differential equations in a desired way. Often there is not enough control to do "complete" inversion, and thus one is led, after performing partial inversion, to the the study of the resulting equations, called the "complimentary dynamics." This area of activity is primarily of engineering nature, and an example will be given to a nonlinear aircraft model. The presentation of this chapter is based on a talk by Blaise Morton from Honeywell on April 1, 1988 and on the technical report [1].

15.1   Linear Systems

The system we want to control has accurate PDE model. However the input actuator degrees of freedom (# of knobs) and the output sensors degrees of freedom (# of dials) convert the problem to ODEs. Let us consider the ODE

$$p(D)X = q(D)U \tag{15.1}$$

where $D = \dfrac{d}{dt}$, U in the control or input, and X is the output. In practice we always have $\deg q \leq \deg p$. Assuming for simplicity that the initial data are zero, we take the Laplace transform $(x(s) = \int\limits_0^\infty e^{-st} X(t)\ dt,\ u(s) = \int\limits_0^\infty e^{-st} U(t)\ dt)$ and obtain

$$p(s)x(s) = q(s)u(s)$$

or

$$x(s) = \frac{q(s)}{p(s)}\ u(s)\ . \tag{15.2}$$

Equation (15.1) (in the time domain) is equivalent to equation (15.2) (in the frequency domain); the rational function

$$g(s) = \frac{q(s)}{p(s)}$$

is called the *transfer function*. The ODE (15.1), for u fixed, is globally asymptotically stable (i.e., every solution tends to zero as $t \to \infty$) if and only if $p(s)$ has no zeros in $Res \geq 0$.

Let us say a few words about filters, since they are going to enter into the subsequent discussion.

A filter in its simplest form is made up of a circuit containing a capacitor with capacitance C and a resistor with resistance R in sequence. The current $I(t)$ then satisfies

$$\frac{dI}{dt} + \frac{I}{CR} = 0$$

so that, for the Laplace transform $i(s)$ of $I(t)$,

$$i(s) = \frac{c}{s + 1/CR}, \qquad c \text{ constant.}$$

By constructing more complicated elements, including also inductors, we obtain for I a differential equation of the form (15.1) (with suitable control parameters embodied in U). Consequently

$$i(s) = \frac{p(s)}{q(s)}.$$

Not every rational function can be achieved by such devices. The engineer will usually have to fulfill some specifications in the frequency domain and he will try to design the filter so as to achieve the specifications as best as possible. One commonly designed filter is a band pass filter that is, given an interval of frequencies $\alpha \le s \le \beta$, the filter is required to be such that the rational function $p(s)/q(s)$ approximates the characteristic function of the interval $\alpha \le s \le \beta$. As another example, if one wants to receive a signal passing through a high frequency noise, then one should introduce a low pass filter, i.e. a filter with $p(s)/q(s)$ small for s large.

A function $g(s)$ holomorphic in $Res \ge 0$ with $|g(\infty)| < \infty$ is said to belong to H_∞ ; RH_∞ consists of the rational functions in H_∞.

Suppose we want to change the dynamics of (15.1) into

$$p_1(D)X = q_1(D)U.$$

We can achieve this by building a prefilter h,

$$h(s) = \frac{p(s)}{q(s)} \frac{q_1(s)}{p_1(s)}$$

in the diagram

$$\xrightarrow{u} \boxed{h} \longrightarrow \boxed{g} \xrightarrow{x}. \tag{15.3}$$

In engineering terms the construction of h is feasible only if

$$\deg q + \deg p_1 \ge \deg p + \deg q_1 \ ;$$

also, right half–plane zeros of p and q cannot be changed. The construction (15.3) is an example of dynamic inversion.

Closely related to the concept of dynamic inversion is the concept of H_∞ optimization. Consider a system (in matrix notation)

$$\dot{X} = AX + BU \ , \tag{15.4}$$
$$Y = CX + DU \tag{15.5}$$

where U is the input, X the state and Y the output or signal, and add a feedback control

$$U = KX \ . \tag{15.6}$$

Given initial values, say $X(0) = 0$, we can solve (15.4)–(15.6) and obtain Y as a function of U. The question is how to design the filter K optimally in terms of the transfer function from U to Y. Going into the frequency domain we have

$$sx = Ax + Bu \ ,$$
$$sy = Cx + Du \ ,$$
$$u = Kx$$

or

$$x = G_{12}u, \quad y = G_{22}u, \quad u = Kx$$

where G_{ij} are matrices of rational functions. More generally consider

$$z = G_{11}w + G_{12}u \ ,$$
$$y = G_{21}w + G_{22}u \ , \tag{15.7}$$
$$u = Ky$$

where w is exogenous input (typically consisting of command signals, disturbances, sensor noises), u the control signal, z the output to be controlled, and y the measured signal; the matrix $G = (G_{ij})$ represents the generalized plant and K represents the control. The block diagram representing (15.7) is given in Figure 15.1

The transfer function from w to z is given by

$$z = \left[G_{11} + G_{12} \, K(I - G_{22}K)^{-1}G_{21} \right] w \ .$$

The standard problem in H_∞ optimization is to find K which minimizes the H_∞-norm of this transfer matrix, under the restriction that K stabilizes G; for details see [2] where a solution to this problem is given; see also [3].

15.2   Nonlinear Systems

We begin with a simple example of dynamic inversion for a nonlinear system. Suppose the real system is $\dot{x} = f(x) + u$, as described in Figure 15.2.

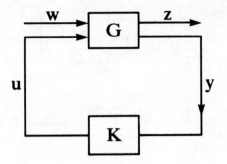

FIGURE 15.1.

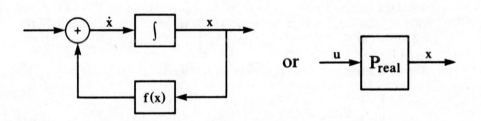

FIGURE 15.2.

and the desired system is $\dot{x} = g(x) + u$. Then we have to build a feedback control $k(x) = g(x) - f(x)$ as described in Figure 15.3.

Usually there are not enough controls for a complete inversion (fewer knobs than dials). In that case we invert the dynamics in a subspace and proceed to analyze the remaining (complimentary) dynamics; we refer to this procedure as *partial dynamic inversion*.

We give a nonlinear aircraft example. The equations have the form

$$f(\dot{x}, x) + g(x)h(x, u) = 0 \qquad (15.8)$$

where f is 6×1, g is 6×4, h is 4×1,

$$x = (\mathbf{V}, \beta, \alpha, \dot{\Psi}, \dot{\Theta}, \dot{\Phi}, \Phi, \Theta),$$
$$u = (T, \delta_a, \delta_e, \delta_r),$$

(U, V, W) is the velocity vector of the *c.g.* (center of gravity) in body axis coordinates, and $(\mathbf{V}, \beta, \alpha)$ is the same in wind–axis coordinates (speed,

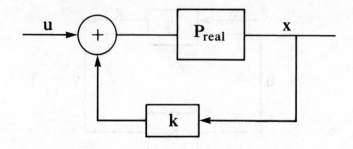

FIGURE 15.3.

sideslip angle, angle of attack), $\mathbf{V} = \{U^2 + V^2 + W^2\}^{1/2}$, and (Ψ, Θ, Φ) are Euler's angles for heading, elevation and bank angle; T is the engine thrust, δ_a the total aileron angle, δ_e the total elevation angle and δ_r the total rudder angle. The system is described in detail in M. Elgersma's thesis [4] and [1].

Let us rewrite (8), more generally, in the forms

$$\begin{bmatrix} \dot{x}_1 \\ \dot{x}_2 \\ \dot{x}_3 \end{bmatrix} = \begin{bmatrix} f_1(x) \\ f_2(x) \\ f_3(x) \end{bmatrix} + \begin{bmatrix} g_1(x) \\ g_2(x) \\ 0 \end{bmatrix} h(x, u) \qquad (15.9)$$

where x is n-dimensional, u is m-dimensional and f is $n \times 1, g$ is $h \times m$, h is $m \times 1$.

We perform dynamic inversion by making

$$\dot{x}_1 = F_1(x, v) \qquad (15.10)$$

and then

$$h(x, u) = [g_1(x)]^{-1}[\dot{x}_1 - f_1(x)] = [g_1(x)]^{-1}[F_1(x, v) - f_1(x)] \qquad (15.11)$$

provided g_1 is invertible. This leads to

$$\begin{aligned} \dot{x}_1 &= F_1(x, v), \\ \dot{x}_2 &= f_2(x) + g_2(x)[g_1(x)]^{-1}[F_1(x, v) - f_1(x)], \qquad (15.12) \\ \dot{x}_3 &= f_3(x). \end{aligned}$$

It now remains to study the complimentary dynamics (15.12). Let us take for example

$$F_1(x, v) \equiv 0 .$$

Then (15.12) reduces to

$$\dot{x}_2 = f_2(x) - g_2(x)g_1^{-1}(x)f_1(x),$$
$$\dot{x}_3 = f_3(x).$$

In the case of the aircraft example this system, after some calculations, takes the form (see [1])

$$\begin{bmatrix} \dot{V} \\ \ddot{\Psi} \end{bmatrix} = \begin{bmatrix} a_{00} + a_{20}V^2 + a_{11}V\dot{\Psi} + a_{02}\dot{\Psi}^2 \\ b_{00} + b_{20}V^2 + b_{11}V\dot{\Psi} + b_{02}\dot{\Psi}^2 \end{bmatrix} \qquad (15.13)$$

or

$$\dot{V} = [1, V, \dot{\Psi}]A \begin{bmatrix} 1 \\ V \\ \dot{\Psi} \end{bmatrix}, \qquad \ddot{\Psi} = [1, V, \dot{\Psi}]B \begin{bmatrix} 1 \\ V \\ \dot{\Psi} \end{bmatrix}$$

where

$$A = \begin{bmatrix} a_{00} & 0 & 0 \\ 0 & a_{20} & \dfrac{a_{11}}{2} \\ 0 & \dfrac{a_{11}}{2} & a_{02} \end{bmatrix}, \qquad B = \begin{bmatrix} b_{00} & 0 & 0 \\ 0 & b_{20} & \dfrac{b_{11}}{2} \\ 0 & \dfrac{b_{11}}{2} & b_{02} \end{bmatrix}.$$

Introducing variables x, y by

$$\begin{bmatrix} f_{11} & f_{12} \\ f_{12} & f_{22} \end{bmatrix} \begin{bmatrix} x \\ y \end{bmatrix} = \begin{bmatrix} V \\ \dot{\Psi} \end{bmatrix} \qquad (f_{ij} \quad \text{constants})$$

we finally obtain (see [1])

$$\begin{bmatrix} \dot{x} \\ \dot{y} \end{bmatrix} = H \begin{bmatrix} 1 - x^2 + ey^2 \\ -2xy \end{bmatrix} \qquad \text{with} \quad e = 1 \quad \text{or} \quad e = -1 \qquad (15.14)$$

provided (15.13) admits real equilibrium points that are nondegenerate (i.e., max (rank A, rank B) = 3) and provided $\lambda A + \mu B$ has rank $\geq 2 \quad \forall \lambda, \mu$.

Theorem 15.1 *[1] If $e = 1$ then $(1, 0)$ and $(-1, 0)$ are the only critical points, and, if $S = \dfrac{1}{2}(H + H^T)$ is positive definite, then*

$$F(z) = |\frac{1 - z}{1 + z}|^2 \qquad (z = x + iy)$$

is a Liapunov function; starting from any (x_0, y_0) with $|F_0| = |F(x_0 + iy_0)| < 1$, there holds

$$|F(z(t))| \leq |F_0|e^{-\lambda t}, \qquad \lambda = \text{min. eigenvalue of } S.$$

Thus $(1,0)$ is (essentially) globally asymptotically stable. Similarly if S is negative definite then $(-1,0)$ is globally asymptotically stable. When $e = -1$,

$$F = x\left(1 - y^2 - \frac{1}{3} x^2\right)$$

is a Liapunov function provided S is positive definite, and then $(1,0)$ is an attractor; the points $(0, \pm 1)$ are saddle points and $(-1,0)$ is repulsive. When S is negative definite the roles of $(1,0)$ and $(-1,0)$ are interchanged.

The above conclusions on stable equilibria can be interpreted in terms of flying qualities. However, it is not clear that every maneuver of the aircraft results in S positive (or negative) definite. Thus the question arises.

Question 1. Find the stability region near a trim (i.e., near an equilibrium) in case S is indefinite.

A more general question is what are the good schemes for dynamic inversion (15.10) in order to achieve the "best" flying qualities. One basic goal is stability in a neighborhood of the maneuvers which the aircraft is designed to execute. A local stability theorem was proved by Sell [5].

15.3 REFERENCES

[1] Nonlinear flying quality parameters based on dynamic inversion, Honeywell SRC, AFWAL –TR–87–3079, October 1987.

[2] B. Francis, *A course in H_∞ Control Theory*, Lecture Notes in Control and Information Sciences, # 88, Springer–Verlag, Berlin, 1987.

[3] C. Foias and A. Tannenbaum, *On the Nehari problem for a certain class of L^∞-functions appearing in control theory*, J. Funct. Anal., 74 (1987), 146–149.

[4] M.R. Elgersma, *Nonlinear Control with Applications to High Angle of Attack and VSTOL Flight*, Master's Thesis, University of Minnesota, 1986.

[5] G. Sell, *Dynamical properties of flight maneuvers*, Appendix B in [1].

16

The Stability of Rapid Stretching Plastic Jets

16.1 Introduction

Shaped charged jet is a fast stretching fluid jet having a nearly cylindrical shape, with possibly some necking, as shown in Figure 16.1.

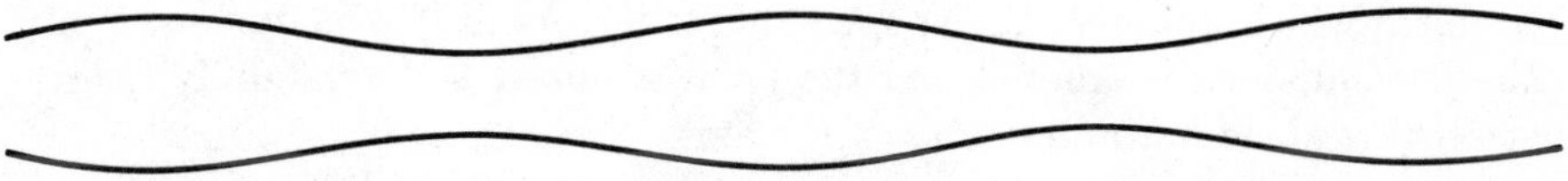

FIGURE 16.1.

A typical shaped charge consists of an explosive with a conical cavity lined up with a thin metal sheet (Figure 16.2). After the initiation of the charge the liner collapses toward the axis where an extremely high velocity jet is formed [1]. The velocity of the jet particles increases with the distance from the rear end, with typical velocities $\sim 8\ km/s$ at the front and $\sim 4\ km/s$ at the rear. The jet experiences enormous stretching during its flight and this increases its penetration capabilities upon impact at a target, since the penetration is proportional to the length of the jet. On the other hand, if the distance from the point of production of the jet to the target is too large, the stretching jet develops instabilities like the necking of bars (see Figure 16.1) and the penetration power is rapidly deteriorating.

Shaped charged jets have been extensively used since World War II in both the military and the civilian environments. Mathematically, the basic question is to understand the instabilities of the jet. Such an analysis was carried out by Frankel and Weihs [2] for a capillary jet of an ideal fluid, assuming that the jet is axially symmetric and stretching in both directions, along its principal axis, with linear velocity; for simplicity the jet is assumed to be of infinite length.

More recently Louis A. Romero from Sandia National Laboratories (Albuquerque, N.M.) has considered the same problem for stretching metal

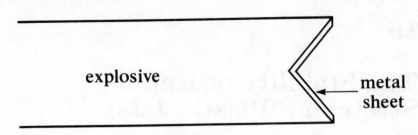

FIGURE 16.2.

jet, which is clearly a more appropriate assumption than that of ideal fluid. Thus the constitutive law is that used for perfectly plastic materials. He presented his results on April 15, 1988, based on his paper [3].

Following [3], we write down in section 16.2 the general free boundary problem which describes the shaped charge jet. As in [2] the dynamics of the jet endpoints is ignored and the jet is assumed to be infinitely long, stretching along both directions of the axis.

In section 16.3 we summarize the stability analysis of Romero, and in section 16.4 some open problems are presented.

16.2   The Free Boundary Problem

It is assumed that the flow obeys the equations for an incompressible perfectly plastic material satisfying the Levy–von Mises equations

$$\rho(\frac{\partial \underline{u}}{\partial t} + \underline{u} \cdot \nabla \underline{u}) = -\nabla p + \text{div } \underline{\underline{T}}' \qquad \text{(momentum equation)}, \qquad (16.1)$$

and

$$\nabla \cdot \underline{u} = 0 \qquad \qquad \text{(equation of continuity)} \qquad (16.2)$$

where ρ is the constant material density, p is the pressure, $\underline{u}$ is the velocity and $\underline{\underline{T}}'$ is the deviatoric stress tensor. In Cartesian coordinates $\underline{\underline{T}}'$ is given by

$$T'_{ij} = 2\mu \dot{\epsilon}_{ij} \qquad (16.3)$$

where $\dot{\epsilon}_{ij}$ is the rate of strain tensor,

$$\dot{\epsilon}_{ij} = \frac{\partial u_i}{\partial x_j} - \frac{\partial u_j}{\partial x_i}, \qquad (16.4)$$

μ is the effective viscosity,

$$\mu = Y(2\dot{\epsilon}_{k\ell} \dot{\epsilon}_{k\ell})^{-1/2} \qquad (16.5)$$

and Y is the yield stress of the material; notice that

$$Y^2 = \frac{1}{2} T'_{k\ell} T'_{k\ell}.$$

Assuming axial symmetry we shall rewrite the above equations in cylindrical coordinates (x, r) with (u, v) the velocity vector:

$$\begin{cases} \rho\left(\dfrac{\partial u}{\partial t} + u\dfrac{\partial u}{\partial x} + v\dfrac{\partial u}{\partial r}\right) = -\dfrac{\partial p}{\partial x} + \dfrac{\partial T'_{xx}}{\partial x} + \dfrac{\partial T'_{rx}}{\partial r} + \dfrac{T'_{rx}}{r}, \\[2ex] \rho\left(\dfrac{\partial v}{\partial t} + u\dfrac{\partial v}{\partial x} + v\dfrac{\partial v}{\partial r}\right) = -\dfrac{\partial p}{\partial r} + \dfrac{\partial T'_{rr}}{\partial r} + \dfrac{\partial T'_{rx}}{\partial x} + \dfrac{T'_{rr} - T'_{\theta\theta}}{r}, \end{cases}$$

$$\tag{16.6}$$

$$\frac{\partial u}{\partial x} + \frac{1}{r}\frac{\partial}{\partial r}(rv) = 0, \tag{16.7}$$

and (16.3) is still valid provided the strain rates are evaluated in cylindrical coordinates:

$$\dot{\epsilon}_{rr} = \frac{\partial v}{\partial r}, \quad \dot{\epsilon}_{\theta\theta} = \frac{v}{r}, \quad \dot{\epsilon}_{xx} = \frac{\partial u}{\partial x},$$

$$\tag{16.8}$$

$$\dot{\epsilon}_{rx} = \frac{1}{2}\left(\frac{\partial u}{\partial r} + \frac{\partial v}{\partial x}\right).$$

We assume stress free boundary conditions at the free boundary $r = R(x, t)$, i.e.,

$$(T'_{rx}, \ T'_{rr} - p) \cdot (-\frac{\partial R}{\partial x}, 1) = 0,$$

$$\tag{16.9}$$

$$(T'_{xx} - p, \ T'_{rx}) \cdot (-\frac{\partial R}{\partial x}, 1) = 0 \quad \text{on} \quad r = R(x, t).$$

Further, to insure that the velocity of the free surface is consistent with the motion of the free surface we must require that

$$\frac{\partial R}{\partial t} + u\frac{\partial R}{\partial x} = v \quad \text{at} \quad r = R(x, t). \tag{16.10}$$

The system (16.6)–(16.10) with (16.3), (16.5) modified in the obvious way (letting (i, j) vary over $(r, r), (x, r), (x, x)$ and (θ, θ)) is the mathematical formulation of the shaped charge jet; naturally one has to incorporate also initial conditions.

Following Romero [3] we shall write down a special solution having the property that the axial velocity of any material point in the jet is assumed to be independent of time and to vary linearly with the distance from the origin. Let (ξ^*, σ^*) be the initial position (in cylindrical coordinates) of a point in the jet, $a_0(0)$ the initial radius of the jet, and $\beta(0)$ the initial strain rate. Then we postulate the velocity vector

$$U_0(\xi^*, \sigma^*, t) = \xi^* \beta(0),$$

$$V_0(\xi^*, \sigma^*, t) = -\frac{1}{2}\beta(0)\sigma^*(q(t))^{-3/2}$$

where

$$q(t) = \beta(0)t + 1 .$$

The position in terms of the Lagrangian coordinates (ξ^*, σ^*) are

$$X(\xi^*, \sigma^*, t) = q(t)\xi^*,$$

$$R(\xi^*, \sigma^*, t) = \sigma^*(q(t))^{-1/2} .$$

In terms of cylindrical Eulerian coordinates (x, r), the velocities are then given by

$$\begin{cases} u_0(x, r, t) = \beta(0)x/q(t), \\ v_0(x, r, t) = -\frac{1}{2}\,\beta(0)r/q(t) . \end{cases} \tag{16.11}$$

The radius of the jet $a_0(t)$ and the axial strain rate $\beta(t)$ are both decreasing functions of time, given by

$$a_0(t) = a_0(0)/(q(t))^{1/2} , \tag{16.12}$$

$$\beta(t) = \beta(0)/q(t) .$$

One can check that if we define the pressure by

$$p_0(x, r, t) = -\frac{3}{8}\rho\beta^2(t)(r^2 - a_0^2(t)) - \frac{Y}{\sqrt{3}} \tag{16.13}$$

then (u_0, v_0, p_0) with $R_0(x, r, t) = a_0(t)$ is a solution of shape charge problem (16.6)–(16.10).

The axial stress integrated over the cross section of the jet is

$$F(t) = \int\limits_0^{a_0(t)} \sigma_{xx}(r)r\,dr = \frac{\sqrt{3}}{2}a_0^2(t)Y\left(1 - \frac{\Gamma^2(t)}{16}\right)$$

where

$$\Gamma^2(t) = \frac{\sqrt{3}}{Y}\,\rho\,a_0^2(t)\beta^2(t) = \frac{\Gamma^2(0)}{q^3(t)} \quad; \tag{16.14}$$

$\Gamma(t)$ measures the relative importance of inertial forces to plastic forces. Note that the jet is in compression as long as $\Gamma^2(t) < 16$ and that $F(t)$ achieves maximum when $\Gamma^2(t) = 4$.

16.3   Stability Analysis

In [3] Romero perturbs

$$u(x, r, t) = u_0(x, r, t) + \delta u(x_i r, t),$$
$$v(x, r, t) = v_0(x, r, t) + \delta v(x, r, t),$$
$$p(x, r, t) = p_0(x, r, t) + \delta p(x, r, t),$$
$$R(x, t) = a_0(t) + \delta a(x, t),$$

and substitutes into (16.6)–(16.10). Setting

$$L(t) = q(t)L(0)$$

where $L(0)$ is the initial axial length scale, suitably chosen, and introducing dimensionless variables

$$s = \log q(t), \quad \xi = \frac{x}{L(t)}, \quad \sigma = \frac{r}{a_0(t)},$$

dimensionless velocities

$$\Phi = \frac{\delta u}{L(t)\beta(t)}, \quad \Psi = \frac{\delta v}{a_0(t)\beta(t)}$$

dimensionless pressure

$$\Lambda = \frac{\delta p}{\rho \, a_0^2(t)\beta^2(t)}$$

and a dimensionless perturbed radius

$$\Omega = \frac{\delta a}{a_0(t)},$$

he derives the equations:

$$\frac{\partial \Phi}{\partial s} + \Phi = -\alpha^2(s)\frac{\partial \Lambda}{\partial \xi} + \frac{1}{\Gamma^2(s)}\left[\frac{1}{\sigma}\frac{\partial}{\partial \sigma}(\sigma\frac{\partial \Phi}{\partial \sigma}) - \alpha^2(s)\frac{\partial^2 \Phi}{\partial \xi^2}\right] \quad (16.15)$$

$$\frac{\partial \Psi}{\partial s} - 2\Psi = -\frac{\partial \Lambda}{\partial \sigma} + \frac{\alpha^2(s)}{\Gamma^2(s)}\frac{\partial^2 \Psi}{\partial \xi^2}, \quad (16.16)$$

$$\frac{\partial \Phi}{\partial \xi} + \frac{1}{\sigma}\frac{\partial}{\partial \sigma}(\sigma\Psi) = 0 \quad (16.17)$$

where

$$\Gamma^2(s) = \Gamma^2(0)e^{-3s}, \quad \alpha^2(s) = \left(\frac{a_0(0)}{L(0)}\right)^2 e^{-3s};$$

the boundary conditions are

$$2\left(\frac{\partial \Psi}{\partial \sigma} + \frac{1}{2}\frac{\partial \Phi}{\partial \xi}\right) - \Gamma^2(s)\wedge + \frac{3\Omega}{4\Gamma^2(s)} = 0 \quad \text{at} \quad \sigma = 1, (16.18)$$

$$\frac{\partial \Phi}{\partial \sigma} + \alpha^2(s)\frac{\partial \Psi}{\partial \xi} = 3\alpha^2(s)\frac{\partial \Omega}{\partial \xi} \quad \text{at} \quad \sigma = 1, \quad (16.19)$$

$$\frac{\partial \Omega}{\partial s} = \Psi \quad \text{at} \quad \sigma = 1. \quad (16.20)$$

To analyze the stability of (16.15)–(16.20) he assumes

$$\Phi = \phi(\sigma, s)e^{i\xi} , \quad \Psi = \psi(\sigma, s)e^{i\xi} ,$$

$$(16.21)$$

$$\wedge = \lambda(\sigma, s)e^{i\xi} , \quad \Omega = w(\sigma, s)e^{i\xi}$$

and substitutes into (16.15)–(16.20) to obtain a system which can be integrated using a finite difference code. We summarize in words the main conclusions derived in [3]:

(i) If $\Gamma(0) \gg 1$, then the jet is initially stable. As the jet stretches, $\Gamma(t)$ decreases until it is small enough ($\Gamma(t) \approx 1$) that the jet becomes unstable. Once the jet has gone unstable, the most unstable wavelength (say $\xi(t)$) is of the same order of magnitude as the radius of the jet (say $r(t)$) at that time, and $\xi(t)/r(t)$ is independent of $\Gamma(0)$.

(ii) If $\Gamma(0) \ll 1$, then the jet is initially unstable. The most unstable wavelength can now be much longer than the radius of the jet; in fact, it is proportional to the radius of the jet times $\Gamma^{-1/5}$. This implies that for slowly stretching jets, the most unstable wavelength can be much larger than the radius of the jet.

The fact that the most unstable wavelength increases as one decreases Γ is in qualitative agreement with experiments of D.E. Grady and D.A. Benson on rapidly stretching metal rings [4].

16.4   Open Problems

Problems.   (1) Study the linear problem (16.15)–(16.20) for existence, uniqueness and asymptotic behavior of solutions; find the shape of $\Omega(x, s)$ (does it indicate necking, is $x \to \Omega(x, s)$ oscillatory?).
   (2)   Consider the free boundary problem (16.5)–(16.10) with initial data

$$R(x, r, 0) = a_0(0), \quad u(x, r, 0) = \frac{\beta(0)x}{q(0)} , \quad v(x, r, 0) = -\frac{1}{2}\frac{\beta(0)}{q(0)}r .$$

Show that the only smooth solution is given by u_0, v_0 (as in (16.11)) and $R(x, r, t) = a_0(t)$.

(3) Take initial values close to the initial values of Problem 2 and prove that there exists a smooth solution for (16.5)–(16.10) for these initial values. Study the shape of the free boundary. (The free boundary problem is of a type similar to that discussed in Chapter 3.)

16.5 REFERENCES

[1] G. Birkhoff, D.P. MacDougall, E.M. Pugh and G.I. Taylor, *Explosives with lined cavities*, J. Appl. Phys., 19 (1948), 563–582.

[2] I. Frankel and D. Weihs, *Stability of a capillary jet with linearly increasing axial velocity (with application to shaped charges)*, J. Fluid Mech., 155 (1985), 289–307.

[3] L.A. Romero, *The instability of rapidly stretching plastic jets*, preprint.

[4] D.E. Grady and D.A. Benson, *Experimental Mechanics*, 23 (1983), 393–400.

17

A Selection of Applied Mathematics Problems

On April 22, 1988 James McKenna from AT&T made a presentation of several problem areas of present interest to AT&T scientists. His talk is reviewed below.

17.1 Path Generation for Robot Cart

Autonomous vehicles for moving materials between workstations are of increasing use in automated factories. There are several methods in use for steering the vehicle. A common method is to paste strips along the desired paths, capable of reflecting light. The vehicle sends light beeps at regular time intervals and detects the reflected light from the strips, using the returned signals for determining its path. This method is effective only in a clean environment; if significant amount of dirt accumulates on the factory floor, the reflecting strip will not function properly. An alternate method consists of laying down a system of underground wires and using magnetic sensing. This however has a serious drawback: when the factory needs to shift from one production line to another, the rebuilding of another underground system of wires is both expensive and time consuming.

A new device was developed at AT&T, a cart (called "Blanch"), moving on three wheels with steering shaft in front and a mounted infra red light source (see Figure 17.1).

The cart cannot make sharp turns; the restriction on the turning angle ω is such that the radii of curvature of the feasable paths are $\geq R_0$, R_0 fixed. Guiding poles are posted in various locations which respond to the light source of the robot. The basic question is:

How to plan the motion of the cart so that it moves from one workstation to another in an optimal way, i.e., taking the shortest collision–free route.

Consider first the extremely simple case where the factory floor is vacant, i.e., the robot is free to move anywhere in $\mathbb{R}^2$. Assume also that the robot is a directed unit segment, i.e., a unit vector. An initial position is given by a point P_0 in $\mathbb{R}^2$ and a direction e, and a target position is similarly given by Q_0, f (Q_0 is any point in $\mathbb{R}^2$ and f is any direction). The problem is: Find a shortest C^1 curve $X = X(s)$ (with s = length parameter, say $0 \leq s \leq L$) such that

$$X(0) = P_0, \quad X'(0) = e \quad \text{and} \quad X(L) = Q_0, \quad X'(L) = f,$$

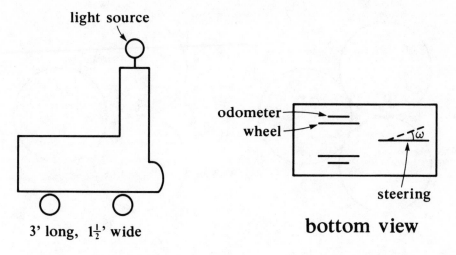

light source

odometer

wheel

steering

bottom view

3' long, $1\frac{1}{2}$' wide

FIGURE 17.1.

$$|X''(s)| \leq \frac{1}{R_0} \quad .$$

This problem was solved by Dubins [1]; he showed that among the optimal paths there is one which consists either of three circular arcs (CCC), or of a circular arc, a straight line segment and a circular arc (CSC); some of the arcs may be absent. Figure 17.2 illustrate a CSC example.

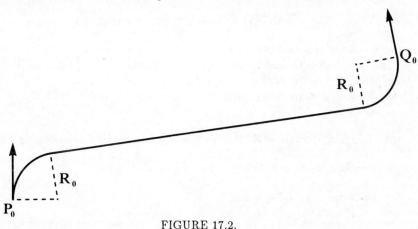

Q_0

R_0

R_0

P_0

FIGURE 17.2.

Recently, J.A. Reed and L.A. Shepp [2] considered the case where the cart is allowed to back-up, i.e, the paths may have cusps (points of reversal). They proved that among the optimal paths one can always find a chain $CCSCC$ with some letters possibly omitted. In Figure 17.3 we give an example with $P_0 = Q_0, e = -f$; in (a) the cart can only go forward and in

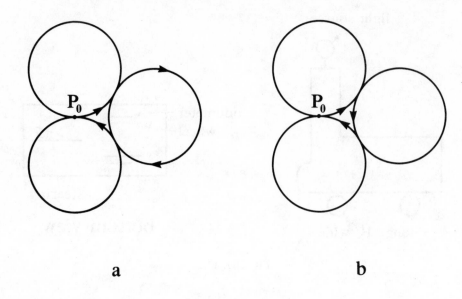

a   b

FIGURE 17.3.

(b) it can back–up.

Problem (1). Study the optimization problem for forward moving cart, under the additional restriction

$$|X''(s) - X''(s')| \le \eta(|s - s'|) \qquad (17.1)$$

where η is a given modulus of continuity.

The motivation for the restriction (17.1) is that engineers do not like an abrupt change in the radius of curvature.

Problem (2). Can the results of Dubins and of Reeds and Shepp be extended to

(a) a bounded domain in $\mathbf{R}^2$ (say a domain bounded by a convex polygon)?

(b) a surface, such as sphere?

(c) the 3–dimensional space $\mathbf{R}^3$?

We now come to the more realistic situation where the factory floor has obstacles. Let us assume for simplicity that all the obstacles are closed polygons, and denote the entire set of these polygons by U. Given (P_0, e) and (Q_0, f), the first question is: are there feasable (i.e. collision–free) paths?

S. Fortune and G. Wilfong [3] constructed a search procedure for a feasable path; complexity of their procedure is exponential: $2^{p(m,n)}$, where n is the number of vertices and m is size of the coordinates of each vertex (i.e., a reduced rational number $\pm p/q$ with $0 \le p, q \le 2^m$) ; $p(m, n)$ is a polynomial in m, n.

In a subsequent work Wilfong [4] considered the case where the vehicle, again with limited steering range, is allowed to move only along m prescribed linear lanes, in a polygonal environment of complexity n. He constructed an algorithm for a motion with minimal number of turns, whose complexity is polynomial in n, m.

Problem (3). Generalize the results of [3] [4] to motions where reverses are allowed.

17.2 Semiconductor Problems

The three basic research areas in semiconductors are processing (or fabrication), devices and circuit. The mathematical theory of device is based on the study of a system of three nonlinear parabolic equations (or elliptic, in the stationary case); some references can be found in Chapter 8. The study of circuit depends on the $V - I$ curves $V = f(I)$ of the devices (cf. Chapter 8 with $I = J$) and involves a large number of coupled devices; mathematically this reduces to a large system of $ODEs$ coupled by means of Kirchoff's law.

Here we shall deal with some problems which arise in processing. The first step is the production of crystals. Typically one inserts a seed crystal, held at the end of a pull rod, into a silicon melt and slowly extracts it from the melt. Under appropriate conditions the crystal grows as it is pulled out; a mathematical description of such a growth (Czochralski growth) is given, for instance, by Langlois [5] [6].

The crystal is sliced into wafers, each used for packing many chips, and each chip is made of a large number of semiconductor devices. Each device consists of an active region of the silicon, isolated from its neighbors, and its electrical properties have to be prepared by doping. We achieve the isolation by oxidation of the silicon in the non–active region to form a silicon oxide (SiO_2) dialectric barrier. To avoid oxidation of the active region, it is masked with a silicon nitride cap impervious to the diffusion of oxygen, which is later removed by etching. The doping process consists of ion implantation of arsenic, for example, which diffuses in the silicon at high temperatures. A simple schematic diagram of the process is shown in Figure 17.4.

This model is described, for instance, in Marcus and Sheng [7] and Chin, Oh, Hu, Dutton and Moll [8]; a mathematical formulation of the free boundary problem was given recently by Tung and Antoniadis [9] and by Tayler and King [10]. The formulation of differential equations in [9] and [10] are the same, but the boundary conditions are different in some important respects.

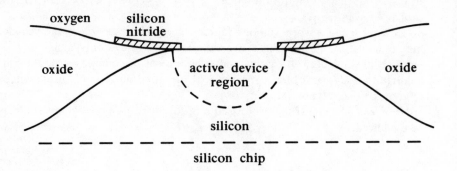

FIGURE 17.4.

Denote by Ω the oxide domain; the boundary consists of three arcs (see Figure 17.5): the oxide–oxygen interface Γ_1, the silicon–nitride cap Γ_2 and the oxide–silicon interface Γ_3.

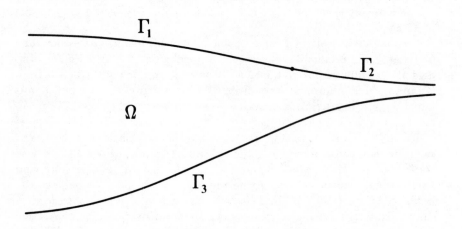

FIGURE 17.5.

Let

$$
\begin{aligned}
c &= \text{concentration of the oxidant,} \\
\mu &= \text{viscosity of the oxide,} \\
\underline{u} &= \text{velocity of the oxide,} \\
P &= \text{pressure of the oxide}
\end{aligned}
$$

In the oxidant transport process, oxidant molecules diffuse from the ambient through the oxide bulk to the oxide–silicon interface to react with silicon. Due to its volume expansion, newly created oxide partly fills the void left by the consumed silicon and partly displaces and pushes existing oxide toward the surface. Thus although initially $\Gamma_1 \cup \Gamma_2$ is a horizontal line, these curves become warped by the transport process.

The above processing takes place at such high temperatures that oxide behaves like a viscous fluid. The equations in the oxide are

$$\nabla^2 c = 0, \tag{17.2}$$

$$\mu \nabla^2 \underline{u} = \nabla P, \tag{17.3}$$

$$\nabla \cdot \underline{u} = 0. \tag{17.4}$$

Introduce the stream function ψ, i.e.,

$$\underline{u} = (-\psi_y, \psi_x).$$

Then the vorticity ω is given by $\omega = -\Delta \psi$, and $\Delta^2 \psi = 0$. Also we have the Cauchy–Riemann equations

$$\frac{\partial \omega}{\partial x} = \frac{\partial (P/\mu)}{\partial y},$$

$$\frac{\partial \omega}{\partial y} = -\frac{\partial (P/\mu)}{\partial x},$$

i.e., P is obtained from ψ by taking the harmonic conjugate of $\mu \Delta \psi$.

We now state the boundary conditions according to [9]: On Γ_1,

$$c = c^*,$$

$$T_{\eta\eta} \equiv -P + 2\mu \frac{\partial u_\eta}{\partial \eta} = 0 \qquad \text{(no stress)}, \tag{17.5}$$

$$T_{\eta\xi} \equiv \mu \left(\frac{\partial u_\eta}{\partial \xi} + \frac{\partial u_\xi}{\partial \eta} \right) = 0 \qquad \text{(no shear)};$$

on Γ_2,

$$\frac{\partial c}{\partial \eta} = 0,$$

$$T_{\eta\eta} = 0, \tag{17.6}$$

$$u_\xi = 0 \qquad \text{(no slip)},$$

and on Γ_3:

$$\frac{\partial c}{\partial \eta} = kc,$$

$$\underline{u} = (\alpha - 1)k_1 c \underline{\eta}, \tag{17.7}$$

$$\underline{v} = k_1 c\eta \quad (\underline{v} = \text{ velocity of the free boundary});$$

here k, k_1 are positive constants, α is the volume of oxide produced from unit volume of silicon and it is equal approximately 2.24, $\hat{\eta}$ is the inward unit normal and $\hat{\xi}$ is the unit tangent vector 90° clockwise from $\hat{\eta}$, and $\dfrac{\partial}{\partial \eta}$, $\dfrac{\partial}{\partial \xi}$ are the derivatives in the directions of $\hat{\eta}$ and $\hat{\xi}$, respectively. The problem (17.2)–(17.7) depends on t only through the last relation in (17.7).

In the above formulation Γ_1 is extended to $+\infty$, Γ_2 to ∞ and Γ_3 to $\pm\infty$; all the arcs Γ_i may be taken as moving or free boundaries. The last condition in (7) provides two scalar equations, with which to determine $\Gamma_1 \cup \Gamma_2$ and Γ_3.

Problem (4). Study the free boundary problem (17.2)–(17.7); find special solutions, prove local existence (in time), etc.

We next formulate a different set of boundary conditions, as given in [9]:
On $\Gamma_1 : y = h(x)$

$$\frac{\partial c}{\partial \eta} = H(c-1) \qquad (H \quad \text{positive constant}),$$

$$\text{no stress, no shear (as in (17.5))}, \qquad\qquad (17.8)$$

$$\frac{\partial h}{\partial t} = -\psi_x - \psi_y \frac{\partial h}{\partial x} \qquad (\Gamma_1 \quad \text{is a material surface}),$$

on $\Gamma_2 : y = \ell(x)$,

$$\frac{\partial c}{\partial \eta} = 0,$$

$$T_{\eta\eta} = -F \frac{\ell''}{(1+\ell'^2)^{3/2}} \qquad (F \quad \text{is a stiffness parameter}), \qquad (17.9)$$

$$\frac{\partial \ell}{\partial t} = -\psi_x - \psi_y \frac{\partial \ell}{\partial x},$$

and on $\Gamma_3 : y = -f(x)$,

$$\frac{\partial c}{\partial \eta} = kc,$$

$$\frac{\partial \psi}{\partial \xi} = \frac{\alpha-1}{\alpha} \frac{\partial c}{\partial \eta}, \quad \frac{\partial \psi}{\partial \eta} = 0, \qquad\qquad (17.10)$$

$$\alpha \frac{\partial f}{\partial t} = c_y + c_x \frac{\partial f}{\partial x}.$$

The first three conditions in (17.10) are the same as the first two conditions in (17.7). The main difference between the boundary conditions (17.5)–(17.7) and (17.8)–(17.10) is that the last condition in (17.7) is replaced by the last condition in (7.9) and the last condition in (17.10).

Some asymptotic analysis considerations are discussed in [10].

Problem (5). Study the free boundary problem (17.2)–(17.4), (17.8)–(17.10).

17.3   Queuing Networks

The simplest queuing model is that of a single server with Poisson arrival time (the probability for n arrival in time $(0,t)$ is $\dfrac{(\lambda t)^n}{n!}\, e^{-\lambda t}$) and exponentially distributed service time T $(P(T < t) = 1 - e^{-\mu t})$. The queue is described schematically in Figure 17.6.

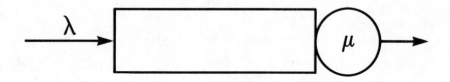

FIGURE 17.6.

If Q is the number of jobs in the queue in service then, provided $\rho \equiv \dfrac{\lambda}{\mu} < 1$,

$$P\{Q = n\} = (1 - \rho)\rho^n \; ;$$

$\rho \approx 1$ indicates congestion.

A more complicated queue, in fact, a queuing network, is described in Figure 17.7.

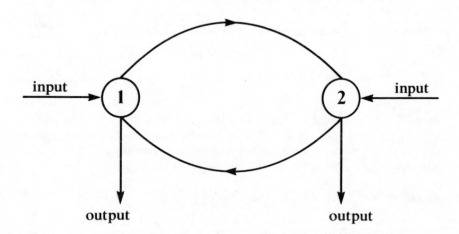

FIGURE 17.7.

with $0 < q_{12} < 1$, $0 < q_{21} < 1$. Customers arrive at station i, and after being served move, with probability q_{12} if $i = 1$ and q_{21} if $i = 2$, to the second station. Denote by $N_i(t)$ the number of customers present at station

i at time t. Assume that the interarrival times are exponentially distributed with mean $1/\lambda_i$ and that the service time at station i is exponentially distributed with mean $1/\mu_i$; Set $q_{11} = q_{22} = 0$, $Q = (q_{ij})$, $\gamma = \lambda(I - Q)^{-1}$, $\rho_i = \gamma_i/\mu_i$. The system is said to be *stable* if $\rho_1 < 1$ and $\rho_2 < 1$.

By a result of Jackson [11]

$$P\{N(t) = n\} = \prod_{i=1}^{2}(1 - \rho_i)\rho_i^{n_i} \ .$$

The case where

$$\alpha = \min\{1 - \rho_1, \ 1 - \rho_2\}$$

is small corresponds to traffic intensities (analogous to $1 - \rho$ small in the case of Figure 17.6 with $\rho = \dfrac{\lambda}{\mu}$). It was proved by Reiman [12] [13] that the process

$$Z_\alpha(t) \equiv \alpha N \left(\frac{t}{\alpha^2}\right) \qquad \text{as} \quad \alpha \to 0$$

is convergent to a reflected Brownian motion in the first quadrant. We shall write down the corresponding elliptic equation and boundary condition for the N-dimensional case in the quadrant $\mathbf{R}_+^N$:

$$\frac{1}{2}\sum_{i,j=1}^{N} A_{ij}\frac{\partial^2 \pi}{\partial x_i \partial x_j} - \sum_{i=1}^{N}\mu_i\frac{\partial \pi}{\partial x_i} = 0 \qquad \text{in} \quad \mathbf{R}_+^N, \qquad (17.11)$$

$$\frac{1}{2}\sum_{j=1}^{N}(2\,A_{ij} - A_{ii}\,R_{ij})\frac{\partial \pi}{\partial x_j} - \mu_i\pi = 0 \qquad \text{on} \quad x_i = 0, \qquad (17.12)$$

$$\pi(x) > 0 \qquad \text{and} \qquad \int_{\mathbf{R}_+^N} \pi(x)\,dx = 1 \qquad (17.13)$$

where (A_{ij}) is a positive definite matrix, $R_{ij} = \delta_{ij} - Q_{ij}$, $Q_{ij} > 0$ if $i \neq j$, $Q_{ii} = 0$ and the spectral radius of (Q_{ij}) is smaller than 1 (this implies that (R_{ij}) is invertible); A_{ij} and Q_{ij} are constants.

It was shown by Harrison and Reiman [14] that (for $N = 2$) there exists a solution of (17.11)–(17.13) of the form $\displaystyle\prod_{i=1}^{2}\alpha_i e^{-\alpha_i x_i}$ if and only if

$$(\mu R^{-1})_i < 0 \qquad \forall\, i \qquad (17.14)$$

and

$$A = \frac{1}{2}\,(\Lambda R + R^T \Lambda) \qquad (17.15)$$

where $\Lambda = (\Lambda_{ij})$, $\Lambda_{ii} = A_{ii}$, $\Lambda_{ij} = 0$ if $i \neq j$; see also [15] for other types of solutions in case (17.15) is not satisfied.

Problem (6). Prove that if (17.14) and (17.15) hold then the solution $\prod_{i=1}^{2} \alpha_i e^{\alpha_i x_i}$ is the unique solution of (17.11)–(17.13). Generalize the results to $N > 2$.

Problem (7). Find necessary and sufficient conditions (on A_{ij}, μ_i, Q_{ij}) for the existence of a solution of (17.11)–(17.13).

17.4 REFERENCES

[1] L.F. Dubins, *On curves of minimal length with a constraint on average curvature and with prescribed initial and terminal positions and tangents,* Amer. J. Math., 79 (1957), 497–516.

[2] J.A. Reeds and L.A. Shepp, *Optimal paths for a car that goes both forward and backwards,* Technical Report, AT&T, 1987.

[3] S. Fortune and G. Wilfong, *Planning constrained motion,* to appear in Proceedings STOC 1988.

[4] G. Wilfong, *Motion planning for an autonomous vehicle,* Technical Report, AT&T, 1988.

[5] W.E. Langlois, *Buoyancy–driven flows in crystal–growth melts,* Ann. Rev. Fluid Mech., 17 (1985), 191–215.

[6] A.D.W. Jones, *Scaling analysis of the flow of a low Prandtl number czochraeski melt,* J. of Crystal Growth, 88 (1988), 465–476.

[7] R.B. Marcus and T.T. Sheng, *The oxidation of shaped silicon surfaces,* J. Electrochem. Soc., 129 (1982), 1278–1282.

[8] D. Chin, S.Y. Oh, S.M. Hu, R.W. Dutton, and J.L. Moll, *Two dimensional oxidation,* IEEE Trans. Electron Dev. ED–30 (1983), 744–749.

[9] T.L. Tung and D.A. Antoniadis, *A boundary integral equation approach to oxidation modelling,* IEEE Trans. Electron Dev., ED–32 (1985), 1954–1959.

[10] A.B. Tayler and J.R. King, *Free boundaries in semiconductor fabrication.* to appear.

[11] J.R. Jackson, *Networks of waiting lines,* Oper. Res., 5 (1975), 518–521.

[12] M.I. Reiman, *Queuing networks in heavy traffic,* Ph.D. Dissertation, Dep. of Operation Research, Stanford University, Stanford, Calif. 1977.

[13] M.I. Reiman, *Open queuing networs in heavy traffic*, Math. Oper. Research, 9 (1984), 441–458.

[14] J.M. Harrison and M.I. Reiman, *On the distribution of multidimensional reflected Brownian motion*, SIAM J. Appl. Math., 41 (1981), 345–161.

[15] G. Foschini, *Equilibria for diffusion models of pairs of communicating computers–symmetric case*, IEEE Trans. Inform. Theory, IT–28, no.2, 1982, 273–284.

18

The Mathematical Treatment of Cavitation in Elastohydrodynamic Lubrication

The lubrication problem described in this chapter was presented by Edward Bissett from General Motors Research Laboratories, on April 29, 1988.

18.1   The Model

In a ball bearing, the balls transfer their applied loads to their containing rings over a small two–dimensional region, separated by a thin layer of lubricant (See Fig. 18.1). The ball is rolling in a fixed direction, say in the positive x-axis, with surface speed u.

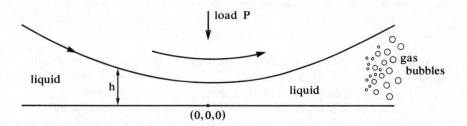

FIGURE 18.1.

Due to the non–slip condition the liquid in contact with the moving surface is also moving with the same velocity. This causes the rest of the liquid to be in motion too. The liquid is assumed to be a Newtonian viscous liquid with viscosity coefficient $\mu = \mu(p)$, where p is the pressure in the liquid. Denote by $h = h(x,y)$ the height of the moving surface and denote by ρ the density of the liquid. The stationary equation governing the pressure distribution in the liquid–film between the ball and the bottom of the liquid was derived from the Navier–Stokes equations by O. Reynolds [1] in 1886:

$$\frac{\partial}{\partial x}\left(\frac{\rho h^3}{12\mu}\frac{\partial p}{\partial x}\right) + \frac{\partial}{\partial y}\left(\frac{\rho h^3}{12\mu}\frac{\partial p}{\partial y}\right) = \frac{\partial}{\partial x}\left(\frac{\rho u h}{2}\right). \tag{18.1}$$

We assume in the sequel that ρ =const. and therefore ρ can be factored out of the equation. Equation (18.1) represents a good approximation for small h (i.e., for film liquid) and it has been confirmed experimentally in regions with non–cavitating films. However cavities do in fact occur; see [2] and the references given there. There are two types of cavities:

(i) *Gaseous cavitation*. If the pressure of the liquid falls below the saturation pressure of the dissolved gases, these gases will be emitted from the solution (see Figure 18.1).

(ii) *Vapour cavitation*. If the pressure falls to the vapour pressure then boiling of the liquid occurs. This may result in erosion of the adjacent surfaces when the vapor bubbles suddenly collapse upon entering a region of higher pressure.

Here we consider only film rupture of type (i). Even in this case the determination of adequate boundary conditions (at the film rupture) is still not completely resolved. The most commonly accepted conditions are the Swift–Stieber conditions (see [2]):

$$p = 0, \quad \frac{\partial p}{\partial x} = 0, \quad \frac{\partial p}{\partial y} = 0 ; \tag{18.2}$$

here we assumed for simplicity that the ambient pressure, which is the same as the pressure in the cavities, is equal to zero. Thus

$$p > 0 \quad \text{in the liquid.} \tag{18.3}$$

Another point which is somewhat ambiguous is the boundary conditions to be imposed on p. It is clear that $p(x,y) \to 0$ if $x^2 + y^2 \to \infty$; however if $x^2 + y^2$ is large enough then $h(x,y)$ is no longer small and the Reynolds equation is thus invalid. The problem is perhaps best considered as a local description that must asymptotically match with an outer flow as $x^2 + y^2 \to \infty$. The boundary "at ∞" then actually represents an overlap domain in which fluid mechanics more complex than that represented by Reynolds equation begins to become important. It seems that a reasonable compromise is to consider the problem (18.1)–(18.5) in a "large" domain Ω and take

$$p = 0 \quad \text{on} \quad \partial\Omega \tag{18.4}$$

So far we have not said anything about the fact that the metal ball may deform under the liquid pressure. For small pressures this deformation may be ignored and thus the problem (18.1)–(18.4) becomes a standard free boundary problem (a "variational inequality"). This formulation has been studied by several mathematicians; see, [3], [4] and the references given there. However, under the substantial loadings typical of practical operating regimes, the ball deforms as an elastic body, and the equations of linear elasticity imply (see [5] or [6]) that

$$h = h_0 + Kp$$

where Kp is a constant times the convolution of the fundamental solution $1/r$ of the Laplace operator with the pressure, and $z = h_0(x, y)$ is the equation of the undeformed lower portion of the solid. Notice that

$$\Delta(h - h_0) = \Delta(Kp) = cp \qquad (c \quad \text{positive constant})$$

which is, in fact, a version of the linear elasticity deformation. Approximating h_0 by $(x^2 + y^2)/2R$ we can rewrite the equation for h in the form

$$h(x,y) = k + \frac{x^2 + y^2}{2R} + \frac{2}{\pi E'} \int\int \frac{p(x', y') \, dx'dy'}{[(x - x')^2 + (y - y')^2]^{1/2}} \qquad (18.5)$$

where E' is the effective elastic modulus and k is some positive constant. The problem (18.1)–(18.5) is an elastohydrodynamic lubrication problem (EHL). The load balance on the ball is

$$P = \int_\Omega p(x, y) \, dxdy.$$

18.2   Roller Bearing

Here, instead of a rolling ball we have a rolling cylinder, and this is precisely the situation of a roller bearing with lubrication. If the cylinder rolls in the positive x-direction with surface velocity u, then the EHL problem is the following:

$$\frac{d}{dx}\left(hu - \frac{h^3}{6\mu}\frac{dp}{dx}\right) = 0, \qquad (18.6)$$

$$h(x) = k + \frac{x^2}{2R} - \frac{4}{\pi E'}\int_{-\infty}^{\infty} p(s) \log|x - s| \, ds \qquad (18.7)$$

and p satisfies (18.2)–(18.4). There has been recent numerical work on this problem; see Bissett and Glander [7] and the references given there. The authors in these papers seek a function $p(x)$ defined for $-\infty < x < \infty$ such that

$$p(x) > 0 \quad if \quad -\infty < x < b, \quad p(x) = 0 \quad if \quad x > b, \quad (18.8)$$
$$p(b) = p'(b) = 0, \qquad (18.9)$$

and p satisfies (18.6), (18.7) for $-\infty < x < b$. Earlier numerical work asserted that $p(x)$ has a sharp spike near $x = b$, but a more careful recent numerical method in [7] shows that the spike has a smooth tip and it decreases to zero at $x = b$ together with its derivative.

The system (18.6)–(18.9) can be written in the form

$$\frac{h^3}{\mu(p)} \frac{dp}{dx} = 6u(h - h(b)),$$ (18.10)

$$h(x) = h(b) + \frac{x^2 - b^2}{2R} - \frac{4}{\pi E'} \int_{-\infty}^{b} p(s) \log \left| \frac{x - s}{b - s} \right| \, ds.$$ (18.11)

Recently E. Bisset [8] has carried out an analysis of this problem for $\mu(p)/\mu(0) = O(1)$ and $u \ll 1$ as a singular perturbation from the corresponding known solution in the absence of lubricant. In this work, reduced forms of (18.10) and (18.11) in the neighborhoods of $x = \pm b$ can be combined into one nonlinear integro–differential equation for h. These subproblems near $x = \pm b$ have been solved numerically by E. Bissett and D. Spence [9]. Work on the more realistic case where

$$\mu(p) = \mu_0 e^{\alpha p}, \qquad \alpha \gg 1$$ (18.12)

is in progress.

18.3   Open Problems

We mention some open problems which came out in a discussion with E. Bissett.

Problems. (1) Prove that (18.6)–(18.9) has a solution; is the solution unique?

(2) Prove that the 2-dimensional problems (18.1)–(18.5) has a solution; is the solution unique?

(3) Describe the shape of the free boundary of the problem (18.1)–(18.5); i.e.,

(i) is $p(x, y) > 0$ for $x < -M$, M large?

(ii) is $p(x, y) = 0$ for $x > N$, N large?

Notice that (18.1)–(18.5) is a variational inequality for p (see [10]) with an elliptic operator whose coefficients depend nonlocally on the unknown function p. In problem (1) we wish not only to solve the variational inequality, but also to prove that the non–coincidence set (i.e., the set where $p > 0$) consists of one (semi–infinite) interval $\{-\infty < x < b\}$, i.e., the free boundary consists of a single point b.

18.4   Partial Solutions

Recently Bei Hu at the Univeristy of Minnesota considered the system (18.1)–(18.5) in case $\mu(p)$ is constant and proved that there exists a solution p which is in $W^{2,\infty}(\Omega)$. Specialized to the case of one–dimension, this implies that $p(x)$ is continuously differentiable near $x = b$, which confirms the calculation of Bissett and Glander [7]. Hu also proved property (3) (i). Work on generalizing these results to the case where the viscosity is given by $\mu_0 e^{\alpha p}$ is in progress.

18.5   REFERENCES

[1] O. Reynolds, *On the theory of lubrication and its application to Mr. Beauchamp Tower's experiments, including an experimental determination of the viscosity of olive oil*, Phil. Trans Roy. Soc. 1886, Series A 177, p. 156.

[2] D. Dowson and C.M. Taylor, *Fundamental aspects of cavity in bearings*, Paper I (iii) in *"Cavitation and Related Phenomena in Lubrication"*, Proc. of 1st Leeds–Lyon Symposium on Tribology, D. Dowson, M. Godet, and C.M. Taylor, eds., 1974. pp. 15–26.

[3] G. Capriz and G. Cimmati, *Free boundary problems in the theory of hydrodynamic lubrication: a survey in "Free Boundary Problems: Theory and Applications"*, vol. II, Pitman, London, 1983, pp. 613–635.

[4] G. Bayada and M. Chambat, *Several aspects of cavitation in lubrication, in "Free Boundary Problems: Applications and Theory"*, vol. IV, Pitman, London, 1985, pp. 365–372.

[5] D. Dowson and G.R. Higginson, *Elastohydrodynamic Lubrication*, Pergamon Press, Oxford, 1966.

[6] S.M. Rohde and K.P. Oh, *A unified treatment of thick and thin film elastohydrodynamic problems by using higher order element methods*, Proc. R. Soc. London, A, 343 (1975), 315–331.

[7] E.J. Bissett and D.W. Glander, *A highly accurate approach that resolves the pressure spike of elastohydrodynamic lubrication*, ASME Journal of Tribology, 110 (1988), 241–246.

[8] E. Bissett, *Asymptotic solution of the elastohydrodynamic lubrication problem for low speed, small viscosity, or small elastic modulus*, to appear.

[9] E. Bissett and D. Spence, *The transition and boundary layers of an asymptotic solution for elastohydrodynamic lubrication of line contacts*, to appear.

[10] A. Friedman, *Variational Principles and Free Boundary Problems*, Wiley & Sons, New York, 1982.

19

Some Problems Associated with Secure Information Flows in Computer Systems

The general problem area of computer security has developed over the past twenty years, but it is only within the last five years that the need for solid mathematical foundations has become apparent. On May 13, 1980 Tom Haigh from Honeywell gave a presentation of recent developments of the subject, including some of his ongoing work. The first part of his talk was a discussion of the threats that are addressed by computer security. The second part of the talk was a description of a more general class of problems associated with access control policy. The write–up below is essentially a written version of his lecture which he sent to us a few days later, with several omissions and additions, and with some input from an article by Boebert, Kain and Young [1].

19.1   Threats and Methods of Response

It is common to identify three classes of threats to the security of computing systems:

(i) Unauthorized disclosure of information;

(ii) Unauthorized modification of data;

(iii) Unauthorized denial of service.

The first one is the most commonly associated with military security, and it is the most studied, but there are still many open questions. In this section we shall deal with this problem area; we shall also consider the more general question of how to control the sharing of system resources so that no data is revealed to unauthorized user.

It is necessary to note that all knowledge one gains about a computer system is gained indirectly; there is no direct observable physical reality. In a typewriter there is a direct and observable mechanical connection between a given keystroke and the appearance of the corresponding letter. In a word processor no such observable link exists. An intermediary, or agent, in the form of a computer program performs the transformation; the action of this agent cannot be directly observed by the user of a machine. Indeed the reliability of the agent is established only by experience. Thus there is no assurance that the program inside the word processor is not performing extra tasks inimical to the interests of the user, such as making clandestine copies, or scanning inputs for indication that a critical document is being prepared and then destroying the document when it is completed; a program of this type is called a Trojan Horse. The Trojan Horse program has two functions. One of the functions is innocent and is visible to the user who invokes the program. The other is hostile and is invisible. The private file is intended to serve as a "back pocket" during the attack.

Since the threats to computer security have to do with controlling accesses to system resources, it seems reasonable to solve the problems by requiring the systems to satisfy security policies that put constraints on the accesses users and their processes have to system entities. These are called *access control policies*.

The most satisfactory early solution to the espionage problem for military systems was the presentation by Bell and La Padula [2]. Their idea was to introduce a lattice of security levels, to partition the system operations into two classes, observe (or read, R) and modify (or write, W), and to identify two classes of system entities:

(a) subjects, which are active entities. These can be thought of as processes, or programs in execution. Each project is assigned a security level, generally the log–in level of the user on whose behalf the program is executing, and

(b) objects, which are passive entities. These can be thought of as containers for data, such as files, buffers, or directories. Each object has an associated security level, which is the least upper bound of the levels of data that may be securely stored in the object.

It is then required that the system satisfy a certain set of properties, including: It is then required that the system satisfy a certain set of properties, including:

If a subject, S, has observe (R) access to an object, 0, then in the lattice of security levels, level (S) dominates level (0)	(SIMPLE SECURITY PROPERTY)	(19.1)

and

If a subject, S, has modified (W) access to an object, 01 and simultaneously has observed access to object, 02, then in the lattice of security levels, level (01) dominates level (S), and	(* PROPERTY)	(19.2)

level (S) dominates level (02).

The reason for the Simple Security Property is fairly obvious; a subject running on behalf of a user who can only log–in at U (unclassified), for instance, should not be able to observe TS (top secret) data, since it is quite likely that the uncleared user will learn the value of the data.

The reason for the * Property is slightly more subtle. It is required, for instance, to control the actions of a Trojan Horse program running at TS. Such a program could legitimately observe data in a TS object and, in the absence of the constraint levied by the * Property, write that the data into an object classified U, which could then be observed by a subject at U. This would result in the circumvention of the intent of the Simple Security Property; TS data would be known in the U world.

In order to accomplish the requirements (19.1), (19.2) we install a reference monitor, which is an access controller described schematically in Figure 19.1.

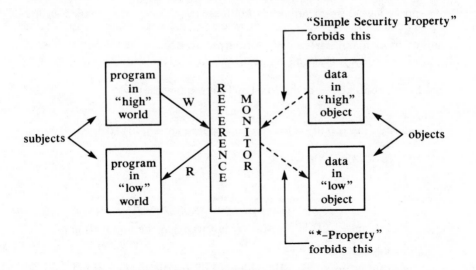

FIGURE 19.1.

This is a subsystem whose task is to check the legitimacy of a program's attempt to access data. A reference monitor is defined to have the following attributes: it must be tamper proof, it must be constructed so that it cannot be bypassed, and it must be small enough and simple enough that it can be thoroughly analyzed and tested.

Although it appears that the rules (19.1), (19.2) impose adequate constraints on the system so that no compromise of high level information is possible, this is actually not the case. Indeed, it is possible to signal information from a high level to a low level without the direct modification of the low level object by the high level subject. This is done by modifying some observable attribute of the system or a system entity. A general discussion of such "covert channels" is given in [3], and an example is given in [4].

Although the fact that the approach of [2] fails to address the covert channel problem was recognized already at the time the approach was developed, it was only in 1982 that a model was developed which addressed itself to this concern. This is the non–interference treatment of security developed by Goguen and Meseguer [5] and elaborated in [6].

In the non–interference model a system is modeled as a state machine, where the inputs are a sequence of instructions obtained by multiplexing instructions issued by subjects in different security contexts and the outputs are generated on a per user, per subject, or per security context basis. Use CON to denote the set of possible security contexts for users and subjects on the system. For the military, or multilevel policy, this would be the set of security levels. In its simplest form a security policy is a binary relation on $CON \times CON, SP$. If (c1,c2) is in SP, then no instruction issued by a subject in c1 may interfere with an output to c2.

More formally, a system satisfies SP from a given initial state, $st0$, if

for all contexts, c, and all instruction sequences, inseq

$$\text{out (int (inseq, } st0), \ c) = \text{out (int (purge (inseq, } c), st0), \ c), \quad (NI)$$

where

> out (st, c) is the output to context, c, from state, st,
>
> int (inseq, st) is the state resulting from the application of inseq to state, and
>
> purge (inseq, c) is the sequence of instructions obtained by purging inseq of all instructions issued by subjects in contexts that should not interfere with c.

For example, in the case of multilevel security, CON is the set of security levels, and the security policy is

$$MLS = \{(L1, L2) : L2 \quad \text{does not dominate } L1\}.$$

That is, if $L2$ is not greater than $L1$ in the lattice of security levels then instructions issued by subjects at level $L1$ should not interfere with outputs to subjects at level $L2$. Thus, if there are only the three security level, $U, S,$

and TS, ordered in the obvious linear fashion, then

$$MLS = \{(S, U), (TS, U), (TS, S)\}.$$

Purge (inseq, U) would be the subsequence of inseq consisting of instructions issued by subjects at U, purge (inseq, S) would be the subsequence of inseq consisting of instructions issued by subjects at U or S, and purge (inseq, TS) would be inseq.

The non-interference model makes good intuitive sense; certainly there cannot be a covert channel from one context to another if subjects in the first context can have no influence on subjects in the second context.

It is easy to show that if a system satisfies MLS from $st0$, then the system satisfies the Simple Security and * Properties in all the states reachable from $st0$ [7].

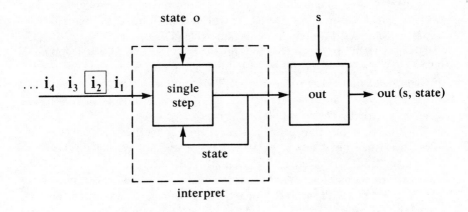

FIGURE 19.2.

The only difficulty with a non-interference model for security is that the proof that a system satisfies such a policy from a given initial state involves a fairly complicated induction over all possible sequences of instructions. For MLS the proof is given in [5]–[7] under some restrictions on the initial states.

There are at least two ways in which the non-interference approach to security modeling needs to be generalized. One is to explore different notions of purgability of instructions. There has been some unpublished work done in this area, but it is too early to say much about the results. The other is to examine the meaning of non–interference in a distributed, and more generally, a non–deterministic, environment. Work on non–deterministic environments has been done by several people including Foley, Jacobs, Mc-Cullough, and Sutherland [8]-[12].

19.2   More General Policies

When the military access control policy is generalized to arbitrary access control policies, connectivity problems arise. To see this, it is convenient to associate a labeled directed graph with each access control matrix. Begin by distinguishing between security contexts for subjects and those for objects. A subject security context will be called a domain and a security center will be called a type. The access control matrix indexed by domains and types will be called a domain definition table (DDT).

Given a DDT, it is possible to define a graph in the following manner. V, the set of vertices, is the set of domains. There is an edge from di to dj if there is a type, t, such that the DDT permits a subject in di to modify objects of type t and the subjects in dj to observe objects of type t. An edge is labeled by the set of all such types.

The first question that arises is that of the connectivity of G. For example, a system is not likely to provide good protection against disclosure of data if the transitive closure of G is complete. On the other hand, it may be that certain connections are more or less desirable while other connections are more or less undesirable. Then the problem is to maximize the utility of the policy (G) by attaching weights to the desirable connections and penalties to the undesirable connections and maximizing an objective function. It is not clear what constitutes an appropriate form of an objective function; this will depend on the specific nature of G.

Let us give an example, based on the purchasing process in a retail store. In this case the domains may be:
- a buyer domain from which the order would be placed,
- a buyer supervisor domain where the order is approved,
- an order writing domain where the actual purchase order is written,
- a receiving domain where the goods are actually received,
- a payment authorization domain where payment for the order is authorized,
- a payment domain where the check is actually written.

In this case it is important that processing occur in precisely this order and that, ordinarily, no one be authorized to act in multiple domains (roles). However, it is also conceivable that, in extraordinary circumstances, someone, say a manager, might be authorized to circumvent some of the procedures. For instance, she may be permitted to authorize the order without going through the intermediate steps. Such an action would be recorded by the systemic audit mechanism.

In general, without knowing the nature of the enterprise, and the details of its security policy, we cannot formulate the precise optimization policy. However, a study of optimization problems of this type may help in the construction and analysis of specific systems. We conclude with formulating one problem of this type.

We are given a finite set of vertices (or domains) v_i. Let $T = \{t\}$ be a set of directed edges, and denote the corresponding digraph by $G(T)$. We associate a penalty $p_{ij} \geq 0$ with each pair of vertices v_i, v_j and define variable x_{ij} by $x_{ij} = 0$ if there is a directed edge in T from v_i to v_j, and $x_{ij} = 0$ otherwise. The total penalty

$$p_1(T) = \Sigma \, p_{ij} x_{ij}$$

is due to disclosure of information. Next we define a reward function. Consider paths, say

$$p = \{v_0, v_1, v_2, \ldots, v_\ell\} \, ,$$

in $G(T)$. If there is no edge in $G(T)$ from v_i to v_j for all $0 \leq i < j \leq \ell$ except when $j = i + 1$ then we write $\epsilon(p) = 1$; otherwise $\epsilon(p) = 0$. This means that paths with redundant domains are going to be discounted. We associate with each path p a reward $\omega(p)$ and introduce the total reward

$$p_2(T) = \Sigma' \omega(p) \epsilon(p) \, ;$$

the prime "$'$" indicates that the summation is restricted to paths p which do not have extensions to paths p' with $p' \neq p, \epsilon(p') = 1$.

We now introduce the functional

$$p(T) = p_2(T) - p_1(T) \, .$$

Problem. Study the graphs $G(T_0)$ for which

$$p(T_0) = \max_T \, p(T) \, ,$$

e.g., find connectivity properties of $G(T_0)$, an algorithm for constructing T_0, etc.

19.3 REFERENCES

[1] W.E. Boebert, R.Y. Kain and W.D. Young, *Secure Computing: The secure Ada target approach*, Scientific Honeyweller, vol. 6, no. 2, July 1985, pp. 1–27.

[2] D.E. Bell and L.J. La Padula, *Secure computer system: Unified composition and multics interpretation*, MTR–2997, Mitre Corp., Bedford, MA, July, 1975.

[3] R.A. Kemmerer, *Shared resource matrix methodology: A practical approach to identifying covert channels*, ACM Trans on Comp. Sys., 1 (1983), 256–277.

[4] J.T. Haigh, R.A. Kemmerer, J. McHugh and W.D. Young, *Experience using two covert channel analysis techniques on a real system design*, IEEE Transactions on Software Engineering, 13 (1987), 157–168.

[5] J.A. Goguen and J. Meseguer, *Security policy and security models*, Proc. Symposium on Security and Privacy, IEEE, 1982, pp. 11–20.

[6] J. Rushby, *Mathematical foundation of the MLS tool for revised special*, Draft internal note, Computer Science Laboratory, SRI International, Menlo Park, Calif., May, 1984.

[7] J.T. Haigh and W.D. young, *Extending the noninterference version of MLS for SAT*, IEEE Transactions on Software Engineering, 13 (1987), 141–150.

[8] D. Sutherland, *A model of information*, Proc. 9th National Security Conference, National Bureau of Standards, Sep. 1986, pp. 175–183.

[9] Daryl McCullough, *Specifications for Multi-Level Security and Hook-up Property*, Proc IEEE Symposium on Security and Privacy, 1987, pp. 161–166.

[10] Daryl McCullough, *Noninterference and the Composability of Security Properties*, Proc IEEE Symposium on Security and Privacy, 1987, pp. 177–186.

[11] Simon Foley, *A Universal Theory of Information Flow*, Proc IEEE Symposium on Security and Privacy, 1987, pp. 116–122.

[12] Jeremy Jacob, *Security Specifications*, Proc IEEE Symposium on Security and Privacy, 1987, pp. 14–23.

20

The Smallest Scale for Incompressible Navier–Stokes Equations

The incompressible Navier–Stokes equations are

$$\frac{\partial \vec{u}}{\partial t} + \vec{u} \cdot \nabla \vec{u} + \nabla p = \nu \Delta \vec{u} ,$$

$$\text{div } \vec{u} = 0$$

(20.1)

where $\vec{u}$ is the velocity vector, p is the pressure and ν is the kinematic viscosity. If we set the viscosity equal to zero, we get the Euler equations

$$\frac{\partial \vec{u}}{\partial t} + \vec{u} \cdot \nabla \vec{u} + \nabla p = 0 ,$$

$$\text{div } \vec{u} = 0 .$$

(20.2)

In order to solve either system one must specify initial and boundary conditions.

It is well known that in two space dimensions both the Navier–Stokes equations and the Euler equations can be solved for all times, say, in a bounded domain, and the solution is unique. However the behavior of the solutions u of the Navier–Stokes equations usually develop turbulent structure for small viscosity. As the viscosity decreases, there is less dissipation of energy in the flow and the size of the smallest features, or scales, diminishes. The relation between the viscosity, the minimum scale and the total energy dissipation is of fundamental interest for understanding of turbulence. A sample of the vast numerical work on this subject is given in references [1]–[6].

Luis G. Reyna from IBM Thomas J. Watson Research Center (Yorktown Heights) has presented on May 20, 1988 some recent and ongoing work jointly with W.D. Henshaw and H.O. Kreiss [7], [8], in which they study the energy dissipation for small ν. This work and some open questions suggested by Reyna are described below.

In [7] [8] the boundary conditions are taken to be periodic of period 2π in each independent space variable, and

$$\vec{u}(x,0) = \vec{u}_0(x)$$

where

$$\nabla \cdot \vec{u}_0\,(x) = 0\,, \qquad \int_Q u_0(x)\,dx = 0$$

(Q = cube of dimension 2π). It is assumed that $\nu \ll 1$ and that

$$\overline{|Du|}_\infty \equiv \sup_{t>0}|D_x u|_\infty \equiv \sup_{t>0}\sup_{x\in Q}|D_x u(x,t)|$$

is finite and $\geq$const.> 0 (uniformly in ν). Assuming also that

$$\max_{0\leq j\leq p}|D_x^j u(x,0)|_{L^2(Q)} \leq C_p \frac{\overline{|Du|}_\infty^p}{\nu^p} \tag{20.3}$$

for all positive integers p and some constants C_p, and writing the solution in the form

$$u(x,t) = \Sigma \hat{u}(\vec{k},t)e^{i\vec{k}\cdot x}\,,$$

they prove that for any $\alpha > 0$ and positive integer j,

$$\sup_{t\geq 0}|\hat{u}(\vec{k},t)|^2 \leq \tilde{C}_j \frac{\overline{|Du|}_\infty^{j+\alpha}}{\nu^{j+\alpha}|\vec{k}|^{2j}} \tag{20.4}$$

where $\tilde{C}_j$ depends only on j and α.

From (20.4) it follows that

$$\sup_{t\geq 0}|\hat{u}(\vec{k},t)|^2 \leq \tilde{C}_j \frac{\overline{|Du|}_\infty^\alpha}{\nu^\alpha}\left\{\left(\frac{\overline{|Du|}_\infty}{\nu}\right)^{1/2}\frac{1}{|\vec{k}|}\right\}^{2j}, \tag{20.5}$$

which implies that the power spectrum $|\hat{u}(\vec{k},t)|^2$ becomes very small if $|\vec{k}|$ is much larger than a constant times $1/\lambda_{\min}$, where

$$\lambda_{\min} = \left(\frac{\nu}{\overline{|Du|}_\infty}\right)^{1/2}; \tag{20.6}$$

i.e., the frequencies which are much larger than $\lambda_{\min}$ have negligible energy. The number $\lambda_{\min}$ is called the *minimum scale* of the flow.

Remark 20.1 The assumption that $\overline{|Du|}_\infty$ is finite is known to hold for two space dimensions; there are no such a priori bound on $\overline{|Du|}_\infty$ in three space dimensions.

In [7] it is proved that the quantity

$$\frac{\nu^{p-1}}{\overline{|Du|}^{p+1}}H_p^2(t) \equiv \frac{\nu^{p-1}}{\overline{|Du|}_\infty^{p+1}}\int_Q|\nabla_x^p u(x,t)|^2\,dx \tag{20.7}$$

remains bounded for all $t > 0$ by its value at $t = 0$ times a positive constant. If the quantity in (20.7) remains bounded from below by a positive constant independent of ν (for a bounded time–interval), then we say that the flow has *maximal dissipation* (in that bounded time–interval).

We define the *energy spectrum* $E(k)$ for wave number k ($0 < k < \infty$) by

$$\hat{E}(k) \equiv \frac{1}{2} \sum_{\||\vec{\ell}|-k|<\frac{1}{2}} |\hat{u}(\vec{\ell})|^2 \ ;$$

then

$$E = \sum_k E(k) \ .$$

By Parseval's identity,

$$H_p^2(t) \sim \int_{k_0}^{1/\lambda_{\min}} k^{2p} E(k) \ dk$$

if k_0 is such that the corresponding integral from $k = 0$ to $k = k_0$ is bounded by a constant times the integral from $k = 0$ to $k = k_0$; recall that for $k > 1/\lambda_{\min}$ $E(k)$ decays exponentially fast in k. It has been conjectured (see [7] and the references given there) that there exists such a k_0 independently of ν and the initial data, and that

$$E(k) \sim k^{-\beta} \quad \text{for} \quad k_0 < k < \frac{1}{\lambda_{\min}} \ . \tag{20.8}$$

If (20.8) holds we call the interval $(k_0, 1/\lambda_{\min})$ the *inertial range*.

If the flow has maximal dissipation and inertial range (in some time interval), then from (20.6)–(20.8), we deduce that $\beta = 3$, i.e.,

$$E(k) \sim k^{-3} \quad \text{for maximal dissipative flows.} \tag{20.9}$$

This was predicted by Batchelor [9] and Kraichnan [10].

Figure 20.1 shows numerical calculations of a flow with $\nu = 10^{-5}$, as it evolves in time. A finite number of vortices develop with some shear layers between them. Furthermore, the flow does not remain maximally dissipative for all times, i.e., $\nu^{p-1} H_p(t)$ is decreasing for large enough times. Thus the energy spectrum steepens, producing a power law

$$E(k) \sim k^{-4} \ ,$$

in agreement with Saffman's theory [11].

For three space dimensions nothing as definite can be said; see Remark above. But even for two dimensions some very basic questions need be studied; some of them are listed below.

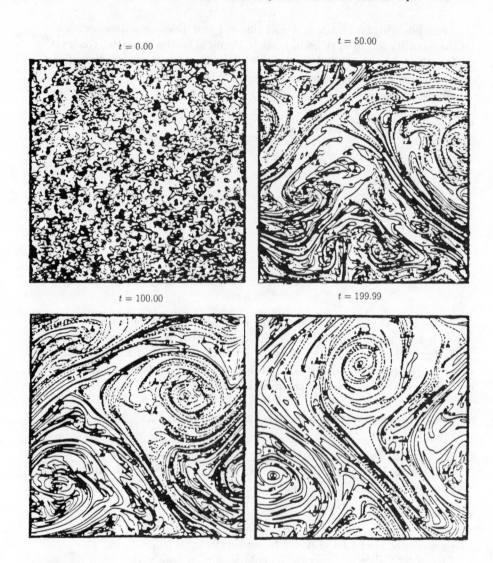

FIGURE 20.1. Contour plots of the vorticity for random initial data, $\nu = 10^{-5}$, $N = 512$.

Problems. (1) For how long does the maximal dissipative stage last? Does this stage reappear?

(2) How are the blobs (in Figure 20.1) formed and how do they move relative to each other?

(3) Is there really an inertial range?

(4) What are the boundary effects, i.e., what is the behavior of a flow with $\nu \ll 1$ in a bounded container?

With respect to problems (1)–(3), there is already substantial numerical work. Thus the interest will be in establishing results by rigorous analysis. In this connection it might be instructive to first consider simplified nonlinear systems exhibiting the same phenomena; for example:

(5) Does Burgers' equation $u_t + uu_x = \nu u_{xx}$ exhibit inertial range? According to Reyna, Burgers' equations exhibits some inertial range, though it is too long compared to what is observed in two–dimensional flows and to what it is believed happens in three-dimensions. In this respect Burgers' equation is a model for compressible flows.

20.1 REFERENCES

[1] S.A. Orszag, *Analytical Theories of Turbulence*, J. Fluid Mech., 41 (1970), 363–386

[2] D.K. Lilly, *Numerical Simulation of Developing and Decaying of Two Dimensional Turbulence*, J. Fluid Mech., 45 (1971), 395–415.

[3] B. Fornberg, *A Numerical Study of 2–D Turbulence*, J. Comp. Phys., 25 (1977), 1–31.

[4] M.E. Brachet, D.I. Meiron, S.a. Orszag, B.G. Nickel, R.H. Morf, U.Frisch, *Small-scale Structure of the Taylor–Green Vortex*, J. Fluid Mech., 130 (1983), 411-452.

[5] J.R. Herring, J.C. McWilliams, *Comparison of Direct Simulation of Two Dimensional Turbulence with Two–Point Closure: the Effects of Intermittency*, J. Fluid Mech., 153 (1985), 229–242.

[6] R. Benzi, G. Paladini, S. Patarnello, P. Santangelo, and A. Vulpiani, *Intermittency and Coherent Structures in Two–Dimensional Turbulence*, J. Phys. A: Math. Gen., 19 (1986), 3771–3784.

[7] W.D. Henshaw, O.H. Kreiss and L.G. Reyna *On the smallest scale for the incompressible Navier–Stokes equations*, Nasa, Langley Research Center, ICASE Report 88–8, Hampton, Virginia, 1988.

[8] W.D. Henshaw, O.H. Kreiss and L.G. Reyna, *On the difference between Euler and Navier–Stokes equations for incompressible flows*, to appear.

[9] G.K. Batchelor, *Computation of the Energy Spectrum in Homogeneous Two-Dimensional Turbulence*, Phys. Fluid. Suppl II (1969), 233–239.

[10] R. Kraichnan, *Inertial Ranges in Two Dimensional Turbulence*, Phys. Fluids, 10 (1967), 1417–1423.

[11] P.G. Saffman *On the Spectrum and Decay of Random Two Dimensional Vorticity Distributions at Large Reynolds Numbers*, Studies Appl. Math., 50 (1971), 377–383.

21

Fundamental Limits to Digital Syncronization

One of the key problems in data transmission is the recovery and maintenance between transmitter and receiver. The unknown delay is exhibited in Figure 21.1.

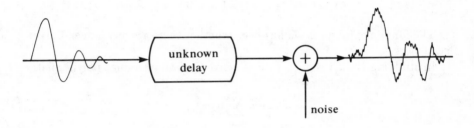

FIGURE 21.1.

In order to syncronize as well as possible special codes are used; notably the Barker code [1]. B. Gopinath from Bell Communications Research presented on June 3, 1988 some recent results of Saffari and Smith regarding the Barker code. The details of his talk and some open problems, which he suggested, are described below.

21.1   The Barker Code

A *Barker sequence* or a *Barker code*, is a finite sequence $\{a_0, a_1, \ldots, a_n\}$ with $a_k = \pm 1$ such that the correlation function $C(\tau) = \sum_{r=0}^{n-\tau} a_r a_{r+\tau}$ satisfies $|C(\tau)| \le 1$ for $\tau \ne 0$. If there is a time delay τ is the transmission of the sequence, then when the receiver matches the received message with the transmitted one, there is a match if $a_i = a_{i+\tau}$ and a clash if $a_i = -a_{i+\tau}$. The correlation $C(\tau)$ is simply the number of matches minus the number of clashes. Therefore the syncronization can best be achieved if the correlation is as small as possible, i.e., if $|C(\tau)| \le 1$.

The known Barker sequences are (see [2]):

length
1	$+1$
2	$+1, +1$
3	$+1, +1, -1$
4	$+1, +1, +1, -1$
5	$+1, +1, +1, -1, +1$
7	$+1, +1, +1, -1, -1, +1, -1$
11	$+1, +1, +1, -1, -1, -1, +1, -1, -1, +1, -1$
13	$+1, +1, +1, +1, +1, -1, -1, +1, +1, -1, +1, -1, +1$

The fundamental question has been whether there are any longer Barker sequences.

In studying Barker sequences it is convenient to introduce polynomials

$$P(z) = \sum_{k=0}^{n} a_k z^k \; ;$$

if $a_k = \pm 1$ then we shall say that P belongs to class Γ_n. Introduce the norms

$$\|P\|_q = \left\{ \int_0^1 |P(e^{2\pi it})|^q \, dt \right\}^{1/q} \; , \quad q > 1,$$

$$\|P\|_\infty = \max_{0 \le t \le 1} |P(e^{2\pi it})| \; .$$

Thus

$$\|P\|_2 = \left\{ \sum_{k=0}^{n} |a_k|^2 \right\}^{1/2} = \sqrt{n+1} \; .$$

Since

$$\|P\|_q \uparrow \quad \text{if} \quad q \uparrow \, ,$$

we deduce that

$$\|P\|_q \ge \sqrt{n+1} \quad \text{if} \quad 2 < q < \infty \, ,$$

and

$$\|P\|_\infty \ge \sqrt{n+1} \; . \tag{21.1}$$

Solving a conjecture of R. Salem, H.S. Shapiro has constructed in 1951, for any $n \ge 1$, a polynomial P in Γ_n such that

$$\|P\|_\infty \le A\sqrt{n+1} \, , \quad A \quad \text{some constant.} \tag{21.2}$$

It is desirable to send signals that have "lots of energy" but not a high peak (so as not to go into the nonlinear region of the device). In terms of the signal associated with P this means that we would like to choose P so that A in (21.2) is as close to 1 as possible. P. Erdös conjectured in 1957 that there is a positive constant C such that

$$\|P\|_\infty \geq (1+C)\sqrt{n+1} \qquad \forall \ \ P \in \Gamma_n \ , \ n = 1,2,\dots \ . \tag{21.3}$$

Recently Saffari and Smith [3] (see also [4]) proved this conjecture and more, namely:

Theorem 21.1 *[4]. There exist $q_0 \in (1, \infty)$ and $C_0 > 0$ such that*

$$\|P\|_{q_0} \geq (1+C_0)\sqrt{n+1} \qquad \forall \ \ P \in \Gamma_n, \quad n = 1,2,\dots \ . \tag{21.4}$$

This implies:

Theorem 21.2 *[4]. There exists a positive constant c_* such that any Barker code must have length $\leq c_*$.*

Proof. Suppose $P(z) = \displaystyle\sum_{k=0}^{n} a_k z^k$ is a Barker polynomial (i.e., $\{a_0, a_1, \dots, a_n\}$ is a Barker sequence). Thus, on $|z| = 1$,

$$P(z)P(\overline{z}) = n+1 + \sum_{\substack{k=-n \\ k \neq 0}}^{n} c_k z^k \quad , \quad |c_k| \leq 1 \ . \tag{21.5}$$

One easily computes that

$$\int_{\{|z|=1\}} |P(z)|^4 = \int_{\{|z|=1\}} |P(z)P(\overline{z})|^2 = (n+1)^2 + 2n$$

since

$$\int_{\{|z|=1\}} z^k = 0 \qquad \text{if} \qquad k \neq 0 \ . \tag{21.6}$$

Similarly, using (21.5) and (21.6), we can get

$$\int_{\{|z|=1\}} |P(z)|^6 = \int_{\{|z|=1\}} |P(z)P(\overline{z})|^3 \leq (n+1)^3 + Cn^2$$

and, more generally,

$$\int_{\{|z|=1\}} |P(z)|^{2q} \leq (n+1)^q + C_q \, n^{q-1} \ .$$

Hence

$$\|P\|_{2q} \leq \sqrt{n+1} \left(1 + O\left(\frac{1}{n}\right)\right) ,$$

which is a contradiction to Theorem 21.1 if $2q > q_0$ and n is sufficiently large, say $n + 1 > c_*$.

Since in practice one needs to transmit lengthy codes, we may wish to relax the Barker property. Let us consider the quantity

$$b_n = \min_{P \in \Gamma_n} \max_{\tau \neq 0} \left| \sum_{k=0}^{n-\tau} a_k\, a_{\tau+k} \right| ;$$

for a Barker sequence $b_n = 1$. If $b_n = O(n^\alpha)$ for some $0 < \alpha < 1$ then the proof of Theorem 21.2 implies that

$$\|P\|_{2q}^{2q} \leq (n+1)^q + O(n^{q-1}\, b_n^q) .$$

Hence if

$$b_n^q = o\left(\frac{1}{n}\right) \qquad \text{for some integer } q \quad \text{with} \quad 2q > q_0$$

then we again conclude that $n \leq c_1$ for some constant c_1. Thus,

$$\text{if} \quad b_n < C\, n^\alpha \qquad \text{for some small} \qquad \alpha > 0 ,$$

$$\tag{21.7}$$

$$\text{then} \quad n \leq \tilde{C} , \quad \tilde{C} \quad \text{a constant depending on} \quad C, \alpha .$$

Problem (1). Find the largest α_0 such that (21.7) holds for all $\alpha < \alpha_0$. (α_0 can be estimated by the q_0 in (21.3); what is the optimal q_0?)

Problem (2). Given $0 < \alpha < \alpha_0$, what is the largest $n = n(\alpha)$ such that there exists a sequence $\{a_0, a_1, \ldots, a_n\}$ with b_n satisfying $b_n \leq n^\alpha$.

It was recently proved by Eliahou, Kervaire and Saffari [5], by arithmetic methods, that there exist no Barker sequences of length larger than 13.

Problem (3). Find the largest n such that there exists a sequence with $b_n \leq 2$; $b_n \leq 3$, etc.

21.2   Complex Barker Sequences

A polynomial $P(z) = \sum_{k=0}^{\infty} a_k\, z^k$ with complex coefficients satisfying $|a_k| = 1$ is called a *unimodular* polynomial. As before,

$$\|P\|_\infty \geq \sqrt{n+1} .$$

Hardy and Littlewood (1930) showed that the unimodular polynomials

$$P(z) = \sum_{k=0}^{n} \exp(i \log k) z^{k}$$

satisfy

$$\|P\|_{\infty} \leq A\sqrt{n+1} \; .$$

Contrary to what happens in the case of real coefficients, J.P. Kahane [6] proved that there are "ultra flat" polynomials P, i.e., for any n there exist polynomials P_n of degree n such that

$$(1 - \delta_n)\sqrt{n} \leq \|P_n\|_{\infty} \leq (1 + \delta_n)\sqrt{n} \qquad \text{with} \qquad \delta_n \to 0 \text{ as } n \to \infty.$$

Thus there is no obstruction here to the existence of arbitrarily long complex Barker sequences. Some examples of complex Barker sequences are given in [2] [7].

Problem (4). Are there complex Barker sequences of arbitrary length?

21.3 REFERENCES

[1] R.H. Barker, *Group synchronizing of binary digital systems*, Communications Theory, Butterworth, London 1953.

[2] S.W. Golomb and R.A. Scholtz, *Generalized Barker sequences*, IEEE Transactions on Information Theory, IT–11, no. 4 (1965), 533–537.

[3] B. Saffari and B. Smith, *Inexistence de polyñomes ultra–plats de Kahane á coefficients* ±1, Preuve de la conjecture d' Erdös, C.R. Acad. Sci. Paris, 306 (1988), 695–698.

[4] B. Saffari and B. Smith, *Une nouvelle méthode d'extrapolation améliorant des inégalités classiques de convexité sur les normes L^p*, C.R. Acad. Sci. Paris, 306 (1988), 651–654.

[5] S. Eliahon, M. Kervaire and B. Saffari, *Closing the list of Barker polynomials in preparation,*

[6] J.–P. Kahane, *Sur les polynomes a coefficients unimodulaires*, Bull, London Math. Soc., 12 (1980), 321–342.

[7] U. Somaini and M.H. Ackroyd, *Uniform complex codes with low autocorrelation sidelobes IEEE Transactions on Information Theory*, September, IT–20 (1974), 689–691.

22

Applications and Modeling of Diffractive Optics

There is an increasing number of diffraction optics applications in industry. J. Allen Cox from Honeywell has described some of the recent applications in his talk of June 10, 1988. His presentation, which is described in the sequel, included a general introduction to diffractive optics. In a subsequent discussion with Allen Cox several open problems were formulated, and these are presented in section 22.3.

22.1 Introduction to Diffractive Optics

Geometrical Optics provides a good approximation whenever the wavelength is small with respect to the geometry of the special features of the surfaces on which the light is incident. In Geometrical Optics electromagnetic waves (i.e., light), represented as rays normal to the wave front, propagate in straight lines in uniform medium. When passing from medium 1 with refraction index n_1 to medium 2 with refraction index n_2, the incident light rays I split into reflected rays R and transmitted rays T, as described in Figure 22.1; n_j is inversely proportional to the speed of light in medium j.

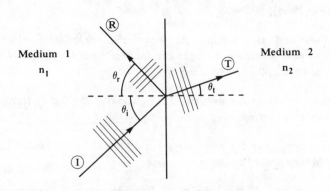

FIGURE 22.1.

The reflection law states: $\theta_r = \theta_i$, and the refraction law (or Snell's law) states that $n_1 \sin \theta_i = n_2 \sin \theta_t$.

Conventional mirrors and lenses are modeled adequately with Geometrical Optics, using ray–tracing methods. Optimal design consists primarily in optimization of surface geometry and material (refractive index) to achieve specified performance requirements, field–of–view, resolution (image quality), etc.

A *diffraction phenomenon* is any deviation of light rays from rectilinear paths which cannot be interpreted by means of reflection and refraction alone [2]. Diffraction phenomena are significant whenever the wavelength λ is of the same order of magnitude as the spatial dimension of surface features (aperture, edges, gratings, etc.). The effect is easily noticeable in acoustics, where λ is small (1 foot to 100 feet).

Diffraction phenomena are described, for instance, in the book of Born and Wolf [1]. For example, Figure 22.2 shows what happens when light rays pass through an array of N slits; the intensity profile is described at the bottom of the figure

Light Rays

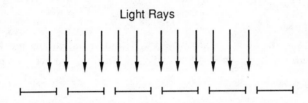

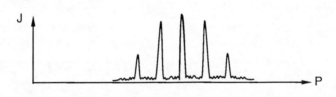

FIGURE 22.2.

Another well known example of optical diffraction is the Fraunhofer diffraction patterns of a circular aperture.

In order to accurately model diffraction effects we have to use the Electromagnetic Vector Field Equations (Maxwell's equations):

$$\nabla \cdot \vec{B} = 0 \,,$$

$$\nabla \cdot \vec{D} = 4\pi\rho \,,$$

$$\nabla \times E + \frac{1}{c}\frac{\partial \vec{B}}{\partial t} = 0 \,,$$

$$\nabla \times \vec{H} - \frac{1}{c}\frac{\partial \vec{D}}{\partial t} = \frac{4\pi}{c}\,\vec{J}$$

where

$$\vec{B} = \text{magnetic induction,}$$
$$\vec{D} = \text{electric displacement,}$$
$$\vec{E} = \text{electric field,}$$
$$\vec{H} = \text{magnetic intensity,}$$

and

$$\vec{B} = \mu \vec{H}, \quad \vec{D} = \epsilon \vec{E}, \quad \vec{J} = \sigma \vec{E} \quad \text{(Ohm's law)};$$

here μ is the magnetic permeability, ϵ is the dielectric constant, σ is the electric conductivity, and $\vec{J}$ is the current density. The conservation law of electrical charge is

$$\nabla \cdot \vec{J} + \frac{\partial \rho}{\partial t} = 0 \, .$$

Across a surface of discontinuity, with normal $\hat{n}_{12}$ going from medium 1 to medium 2, the weak form of the Maxwell equations yield the jump relations:

$$\hat{n}_{12} \cdot \left(\vec{B}^{(2)} - \vec{B}^{(1)} \right) = 0 \, ,$$

$$\hat{n}_{12} \cdot \left(\vec{D}^{(2)} - \vec{D}^{(1)} \right) = 4\pi\sigma \, , \qquad \sigma = \text{surface charge density}$$

$$\hat{n}_{12} \times \left(\vec{E}^{(2)} - \vec{E}^{(1)} \right) = 0 \, ,$$

$$\hat{n}_{12} \times \left(\vec{H}^{(2)} - \vec{H}^{(1)} \right) = \frac{4\pi}{c} \vec{k} \, , \qquad \vec{k} = \text{surface current density}.$$

In a source–free region $\vec{J} = 0$, $\rho = 0$, so that

$$\nabla \cdot (\mu \vec{H}) = 0 \, , \quad \nabla \cdot (\epsilon E) = 0$$

$$\nabla \times \vec{E} + \frac{\mu}{c} \frac{\partial \vec{H}}{\partial t} = 0 \, , \quad \nabla \times \vec{H} - \frac{\epsilon}{c} \frac{\partial \vec{E}}{\partial t} = 0 \, ,$$

from which we easily obtain the wave equations

$$\nabla^2 \vec{E} - \frac{\epsilon\mu}{c^2} \frac{\partial^2 \vec{E}}{\partial t^2} + \nabla(\log\mu) \times \nabla \times \vec{E} + \nabla(\vec{E} \cdot \nabla \, \log\epsilon) = 0 \, ,$$

$$\nabla^2 \vec{H} - \frac{\epsilon\mu}{c^2} \frac{\partial^2 \vec{H}}{\partial t^2} + \nabla(\log\epsilon) \times \nabla \times \vec{H} + \nabla(\vec{H} \cdot \log\mu) = 0 \, .$$

Finally, in homogeneous material,

$$\nabla^2 \vec{E} - \frac{\epsilon\mu}{c^2}\frac{\partial^2 \vec{E}}{\partial t^2} = 0 \,,$$

$$\nabla^2 \vec{H} - \frac{\epsilon\mu}{c^2}\frac{\partial^2 \vec{H}}{\partial t^2} = 0 \,,$$

22.2 Practical Applications

We describe a schematic way to construct a diffraction grating which replaces a conventional lens. Figure 22.3 depicts a conventional lens which focuses at Q all the light rays coming from a source at P and incident to the lens.

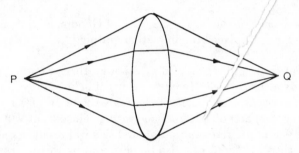

FIGURE 22.3.

In Figure 22.4 we place sources both at P and at Q; the light rays arrive at a holographic plate and interference fringes appear. These are developed into a picture which is used to produce a diffraction grating. If we now remove the source at Q and replace the holographic plate by the diffraction grating, then the rays from P after passing through the grating will focus at Q.

Honeywell has developed software which does the replacement of a conventional lens by diffraction grating without going through the physical experiment with the holographic plate. The software provides a cheap way of manufacturing diffractive grating. There are a number of application areas where a complex and expensive set of optical lenses can be conveniently replaced by diffraction gratings:

(1) A sequence of lenses can be used to transfer gauge readings located in the cockpit of a helicopter so that the pilot can see the readings in front view on the transparent helmet. Some of the lenses are heavy or expensive. It is desired to replace them with diffraction gratings.

(2) Computation of achromatic Fourier transform, or, equivalently, computation of the intensities of the various wavelengths, can be done

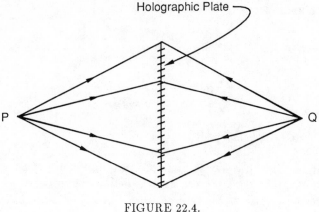

FIGURE 22.4.

by a sequence of lenses; see the book of Goodman [2]. Some of these lenses can be replaced by diffraction gratings.

An unconventional application of diffraction grating to optical computing is under study at Honeywell. Here two beams of light coming from different angles will fall on a planar surface of a body made up of waveguide (i.e. resonant structure) and of a diffraction grating. The output will depend on whether the beams are resonant or not, and this will result in operations of "or" and "either". If "not" can also be produced by this instrumentation, then all the optical logic circuit concepts can be implemented.

22.3 Mathematical Modeling

The problem is to accurately predict the field intensity distribution, reflected and transmitted, by a two dimensional periodic surface structure (grating), such as in Figures 22.5, 22.6.

We make the simplifying assumptions:

The solutions are time harmonic waves: $\vec{E}(\vec{r},t) = \vec{E}(\vec{r})e^{-i\omega t}$,

the region is source–free: $\vec{J} = 0, \rho = 0$;

no magnetic material: $\mu = 1$, and

the incident plane waves are either $\vec{E}(\vec{r}) = \vec{A}\, e^{i\vec{k}\cdot\vec{r}}$ or

$\vec{H}(\vec{r}) = \vec{B}\, e^{i\vec{k}\cdot\vec{r}}$.

The field equations are

$$\nabla \cdot \vec{H} = 0 , \tag{22.1}$$

Medium 1

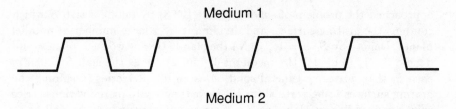

Medium 2

FIGURE 22.5.

Medium 1

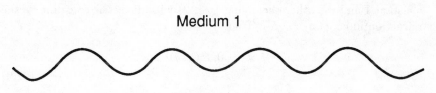

Medium 2

FIGURE 22.6.

$$\nabla \cdot (\epsilon \, \vec{E}) = 0 \, , \tag{22.2}$$

$$\nabla \times \vec{E} - i\frac{\omega}{c} \, \vec{H} = 0 \, , \tag{22.3}$$

$$\nabla \times \vec{H} + i\frac{\epsilon}{c} \, \vec{E} = 0 \, . \tag{22.4}$$

The boundary conditions on the interface are

$$\hat{n}_{12} \cdot (\vec{H}^{(2)} - \vec{H}^{(1)}) = 0 \, , \tag{22.5}$$

$$\hat{n}_{12} \cdot (\epsilon_2 \, \vec{E}^{(2)} - \epsilon_1 \, \vec{E}^{(1)}) = 0 \, , \tag{22.6}$$

$$\hat{n}_{12} \times (\vec{E}^{(2)} - \vec{E}^{(1)}) = 0 \, , \tag{22.7}$$

$$\hat{n}_{12} \times (\vec{H}^{(2)} - \vec{H}^{(1)} = 0 \tag{22.8}$$

Thus, across the interface, $\vec{H}$ is continuous, the tangential components of $\vec{E}$ are continuous, and the normal component of $\epsilon \, \vec{E}$ is continuous.

An introduction to the mathematical formulation of diffractive gratings can be found in the book [3]. More recently Gaylord and Moharam [4] have approached the problem of solving (22.1)–(22.8) by taking a slab S which contains the grating surface, and dividing it by a large number of parallel planes. Denote by S_i $(i = 1, \ldots, N)$ the slab between two such planes, and denote by Π_1 (and Π_2) the mediums 1 (and 2) minus the slab S. Then in each S_i they write the general solution as an infinite series (assuming the grating surfaces to be vertical in S_i). Next they try to fit the various series solutions in adjacent slabs, across their common boundary, as well as to fit the solutions in S_1 and S_N with the solutions in Π_1 and Π_2 across the common boundary. The method is reminiscent of the method of superposition indicated in Chapter 1, although the situation here is much more complicated, especially since N is large. Thus it would be very important to develop mathematical ideas and tools to analyze the problem.

To start the process we suggest two specific and somewhat simple problems.

Suppose there are two media: one occupying $\{y > 0\}$ with dielectric constant 1 and the other occupying $\{y < 0\}$ with dielectric constant ϵ_2; the refraction index is $n_2 = \sqrt{\epsilon_2}$. Then

$$\vec{H} = (0, 0, H(x, y)) \, ,$$

$$\vec{E} = \frac{c}{\epsilon i} \left(H_y, -H_x, 0 \right)$$

(22.9)

is a solution of (22.1)–(22.4) in $\{y > 0\}$ provided

$$H(x, y) = e^{i(\alpha x - \beta^{(1)} y)}$$

where $\alpha^2 + \beta^{(1)^2} = \omega/c^2$. We try to find the refracted and transmitted electromagnetic waves in the same form (22.9), with

$$H(x, y) = e^{i(\alpha x - \beta^{(1)} y)} + r(\alpha) e^{i(\alpha x + \beta y)}, \quad y > 0$$

(22.10)

$$= t(\alpha) e^{i(\alpha x - \beta^{(2)} y)} \, , \quad y < 0 \, .$$

The Maxwell equations then hold $\{$ in $y > 0\}$ if and only if $\beta = \beta^{(1)}$ (the reflection law) and, in $\{y < 0\}$, if and only if

$$\beta^{(2)} = \sqrt{k_0^2 n_2^2 - \alpha^2} \quad \text{(the refraction law)}$$

(22.11)

where $\beta^{(1)} = k_0 \cos \theta$, $\alpha = k_0 \sin \theta$; k_0 is the wave number $2\pi/\lambda_0$, $\lambda_0 = $ wavelength.

Next, the jump relations (22.5), (22.7) are trivially satisfied, whereas (22.6) and (22.8) lead to

$$1 + r(\alpha) = t(\alpha) \, ,$$
$$n_2^2 \beta^{(1)} (1 - r(\alpha)) = \beta^{(2)} t(\alpha) \, ,$$

or, equivalently, to the Fresnel formulas,

$$r(\alpha) = (n_2^2 \beta^{(1)} - \beta^{(2)})/(n_2^2 \beta^{(1)} + \beta^{(2)}),$$

$$\text{(22.12)}$$

$$t(\alpha) = 2n_2^2 \beta^{(1)}/(n_2^2 \beta^{(1)} + \beta^{(2)}) \, .$$

Suppose now that the surface separating the two media is given by a periodic diffraction grating $y = f(x)$ of period L, and

$$f(x) = \begin{cases} 1 & \text{if } 0 < x < \frac{L}{2} \\ -1 & \text{if } \frac{L}{2} < x < L \, . \end{cases}$$

Problem (1). Prove that there exists a unique bounded solution to the Maxwell equations, say $(H_1, H_2, H_3), (E_1, E_2, E_3)$ and that if $L < 2\pi/k_0$ then

$$\||H_1|^2 - (1 + r^2(\alpha) + 2r(\alpha) \int_0^1 \cos(2\beta^{(1)}(\cdots \cdots)a\sigma(t))| + H_2^2 + H_3^2 \to 0$$

$$\text{if } y \to \infty \, ,$$

$$|H_1^2 - t^2(\alpha)| + H_2^2 + H_3^2 \to 0 \quad \text{if } y \to -\infty$$

where σ is some measure.

The assertion of Problem 1 is actually a conjecture based on an analogous situation established at the end of Chapter 1.

Consider next the case where the diffraction grating is very thin, and is given by

$$y = \delta h(x) \, , \qquad \delta \text{ positive and small.}$$

Problem (2). Prove that there exists a unique bounded solution $(H_1^\delta, H_2^\delta, H_3^\delta), (E_1^\delta, E_2^\delta, E_3^\delta)$ of the Maxwell equations, and that

$$H_1^\delta(x, y) - H(x, y) = \delta K(x, y) + o(\delta)$$
$$H_2^\delta = o(1), \ H_3^\delta = o(1) \qquad \text{as } \delta \to 0$$

where $H(x, y)$ is the solution corresponding to $\delta = 0$, given by (22.10).

22.4   References

[1] M. Born and E. Wolf, *Principles of Optics*, Pergamon Press, sixth Edition, Oxford, 1980.

[2] J.W. Goodman, *Introduction to Fourier Optics*, McGraw Hill, New York, 1968.

[3] *Electromagnetic Theory of Gratings*, R. Petit, editor, Springer Verlag, Berlin, 1980.

[4] T.K. Gaylord and M.G. Moharam, *Analysis and applications of optical diffraction by gratings*, Proceedings IEEE, 73 (1985), 894–937.

Index